Ultrastructure of the Root-Soil Interface

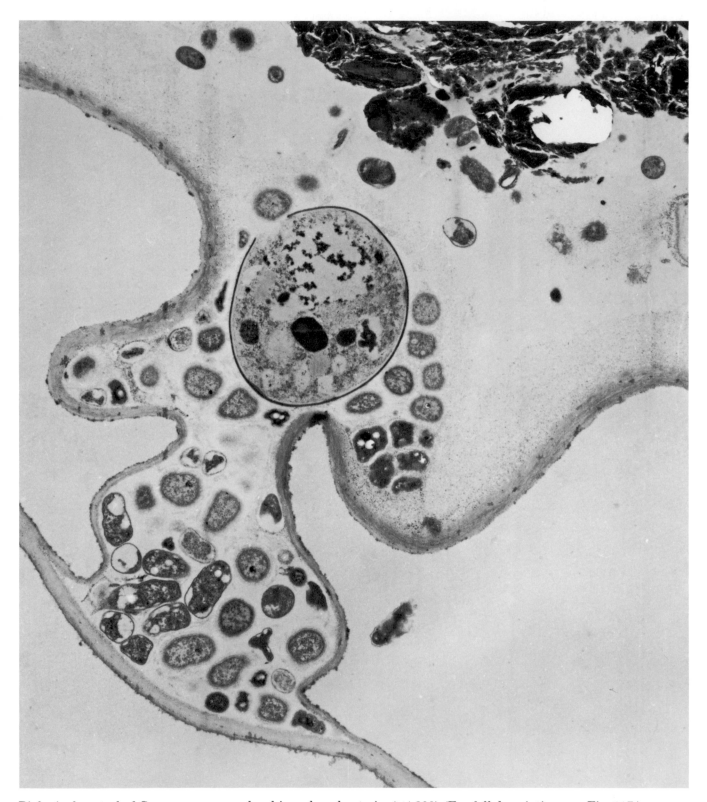

Biological control of *Gaeumannomyces* by rhizosphere bacteria. (×4,300) (For full description, see Fig. 117.)

Ultrastructure of the Root-Soil Interface

R. C. Foster, A. D. Rovira, and T. W. Cock

CSIRO Division of Soils
Adelaide, South Australia

The American Phytopathological Society
St. Paul, Minnesota, U.S.A.

Library of Congress Catalog Card Number: 83-70968
International Standard Book Number: 0-89054-051-9

Printed in the United States of America

The American Phytopathological Society
3340 Pilot Knob Road
St. Paul, Minnesota 55121, USA

Foreword

Few subjects related to the growth and development of higher plants attract the interest of as many separate disciplines as does the subject of the root-soil interface. K. F. Baker and I, in our first book, *Biological Control of Plant Pathogens*, recognized this about the rhizosphere when we wrote:

> To the soil microbiologist, the rhizosphere is the narrow soil zone surrounding living plant roots, which contains root exudates, sloughed root remains, and large populations of microorganisms of various nutritional groupings. To the plant physiologist it is the zone of ion uptake and exchange, of oxygen and carbon dioxide exchange, and of the mucigel matrix. To the soil physicist it is the zone of minimal porosity, of water diffusion and uptake, and of water-potential gradients. To the plant pathologist it is the zone where root pathogens are stimulated by root exudates, and where they swarm, grow ectotrophically, or form infection structures prior to pathogenesis.

To this list can be added the structuralist concerned with the ontogeny and fate of outer root cells, especially the outer cell wall materials, and the ecologist concerned with ecological niches and successions of organisms. To each of these areas of scientific endeavor, the root-soil interface presents mystery, fascination, and research challenges as important to mankind as any in the physical and biological sciences today.

By bringing together in a literary way the 119 plates of high quality electron micrographs of the root-soil interface, selected from hundreds of electron micrographs they have made during the past 10–15 years, R. C. Foster, A. D. Rovira, and T. W. Cock have produced a book that serves the needs of all disciplines concerned with roots and with the soil surrounding the roots. All mystery about this important part of the plant and its environment can be attributed, to one degree or another, to the fact that it cannot be seen, or can only be abstracted from models and indirect measurements. Remarkably, the ultrastructural views in this book are of root surfaces and rhizosphere soil taken intact from the field and processed in a way that left the root-soil interface essentially undisturbed. By so doing, the authors have demonstrated their unique talents as artists and scientists.

The material presented in this book is concerned with wheat, paspalum, ryegrass, clover, rape, chickpea, rice, and pine. These kinds of plants—monocots and dicots of the angiosperms and one member of the gymnosperms—represent all major categories of vascular plants. Moreover, by combining the latest methodology in histochemistry with both transmission and scanning electron microscopy, the authors have provided photographs in this book that not only illustrate the structure of the root-soil interface but also reveal much about function and even dynamics of the rhizosphere. A particularly significant scientific advance, made possible by their histochemical-ultrastructural approach, is the demonstration that the mucilage commonly observed to overlay the root surface in the regions of elongation and maturation is actually the remains of the outer primary wall of root epidermal cells and is not special secretions from underlying epidermal or cortical cells as once suspected. Their approach also reveals, embedded in the mucilage, microorganisms heretofore undescribed. Like any scientific contribution of merit, the findings presented in this treatise answer some questions but raise many others about the root-soil interface.

This book may ultimately prove most useful to those working in the more applied areas of rhizosphere microbiology, such as on symbiotic and associative nitrogen fixation, mycorrhizas, and biological control of root pathogens. Each of these research areas faces the challenge of managing populations of resident or introduced microorganisms in the rhizosphere. The photographs in this book provide us with a view of the spatial, structural, and even temporal relationships of the sites where these beneficial microorganisms must become established and function. Many of the photographs capture various stages of biological control or of symbiosis. The manipulation and enhancement of desirable microbiological processes at the root-soil interface, through conventional approaches or by new biotechnology, as a means to increase the supply of mineral nutrients to roots and to keep the roots healthy, are among the most important yet most underexploited areas in crop production today. Goals such as increasing the efficiency of crop production, reducing the dependence of agriculture on fossil-fuel energy, and pushing yields of important food crop plants above the so-called yield plateau cannot be achieved without the right breakthroughs in applied rhizosphere microbiology. This book is a landmark contribution to rhizosphere microbiology. It will serve as a resource and will greatly enlighten workers in this field.

R. James Cook

Preface

The electron micrographs presented in this book demonstrate the dynamic nature of the root surface in space and time as root surface cells differentiate, function, and senesce. The micrographs also illustrate the wide diversity of microorganisms in the rhizosphere, resulting from the morphological and biochemical changes in the root-soil interface.

It is only in the last 10 years that techniques for preparing soils for electron microscopy have been reliable enough to attempt a detailed ultrastructural study of the rhizosphere and that cytochemical methods have been sufficiently specific to add useful information about the chemical composition of rhizosphere components of submicron size.

Because of the limitations on space in scientific books and periodicals and the cost of reproducing high quality photographs, electron microscopists in biological fields are only able to publish a very small proportion (<1%) of the electron micrographs they produce. The necessity to trim electron micrographs to fit the usually small formats allowed by the normal channels of publication is a further frustration. This means that extraneous detail, however interesting and informative, must be removed, so that most published work has consisted of small portions of electron micrographs at high magnification to illustrate particular details. This is unfortunate, because the ultrastructure of soils and rhizospheres is particularly heterogeneous, and broad formats are needed to give the overall views necessary for proper understanding of the complexity of rhizosphere ultrastructure.

Fortunately, the need for collections of large format electron micrographs on topics of general or specific biological interest has been recognized for some time, and in this book we present some of the data we have collected in a much larger format than is usually allowed. Where possible, we have included large areas at low magnification so that some of the variability in soils and at root surfaces can be clearly seen. Our aim is to enable readers to visualize the relative positions of the components of the soil and the root-soil interface in a three-dimensional form. We hope that such a conceptual approach to this complex environment will help develop basic principles of value in the fields of root growth, plant nutrition, root pathology, and the biological control of root diseases.

Where possible, we have also attempted to display scanning and transmission electron micrographs of similar materials opposite one another so that the complementary nature of the two techniques in rhizosphere investigations can be fully appreciated. We hope that this book will give students in soil science, plant nutrition, and plant pathology, as well as others, a much broader perspective of the root-soil interface than is possible by reading the normal scientific literature.

The scanning electron micrographs in the book are from a cooperative project with R. Campbell, University of Bristol, and were printed by J. A. Coppi. We are also grateful to Dr. Campbell for allowing us to print some of his other scanning electron micrographs (Figures 100, 101, and 110).

We thank Mrs. P. Udompongsanon and especially Miss Y. K. McEwan for excellent technical assistance in preparing difficult materials for transmission electron microscopy. The line drawings were prepared by G. E. Rinder and R. M. Schuster of CSIRO. Miss McEwan also rendered invaluable help in preparing the text for publication.

Contents

Introduction

Of all natural materials, soils are perhaps the most heterogeneous, consisting of minerals, organic matter, water, salts in solution, gases, roots of plants, small animals, and microorganisms.

Except for some unusual soils such as peats, minerals are the major component of soils and are derived by chemical and physical breakdown of the underlying rocks. As in other ecosystems, all of the organic matter of soils originates from green plants, consisting mainly of cell wall fragments from roots, leaves, and stems and of mucilages and soluble organic compounds released from roots. These residues are comminuted by the action of insects, mites, and other soil animals, but soil microorganisms (bacteria, actinomycetes, and fungi) are mainly responsible for their decomposition and transformation into humic materials. Besides breaking down soil organic matter, bacteria and fungi secrete carbohydrate slimes and gels. Such polysaccharides constitute about 15–20% of the soil organic matter and hold the soil components together to form microscopic aggregates that, in turn, are loosely bound by the hyphae of fungi and actinomycetes to form soil crumbs. Thus the characteristic properties responsible for the physical stability and chemical fertility of soils are due to the activities of the microorganisms and the soil fauna.

Of the organic matter in the soil at the end of these biological processes, 80% is humic material, made of complex, stable polymers of carbohydrates, polyphenols, and amino acids. The morphology of humic materials depends on the conditions under which they formed in the soil, but they consist of fibers or globular deposits about 10 nm in diameter. Carbon dating shows that some humic materials that are complexed with metal ions from the soil may be more than 1,000 years old.

Land plants depend on the soil for both water and nutrients: the amount of root produced by a plant may be as great as and even larger than the aerial parts, especially under dry conditions. A 16-week-old winter rye plant produces 13×10^6 root axes and laterals with a total length of more than 50 km and a surface area exceeding 200 m². Despite this, most roots are so fine that in many soils they occupy only a small volume. At flowering, roots of winter wheat occupy 1% of the volume of the top 15 cm of soil. Roots seldom occupy more than 5% of the soil volume (Russell, 1977).

The mass flow of water to the root surface engendered by the transpiration stream moves soluble nutrients such as nitrate to the root, but roots must continually explore new soil to obtain less soluble nutrients such as phosphate and zinc. This is one reason why plants must constantly produce new roots. In one day, each root tip may produce 18,000 new cells and extend up to 5 cm. When the nearby soil has been denuded of insoluble nutrients, autolysis of the redundant outer tissues responsible for nutrient uptake occurs. This means that the parts of the root that release readily available organic substrates age rapidly, and at first, only bacteria that can grow and multiply quickly will accumulate in the rhizosphere. When the low molecular weight substances no longer escape from the root, these bacteria are replaced by slower growing microorganisms such as fungi and actinomycetes, which can use complex substrates such as cell wall materials.

The boundary between the root and the soil changes constantly because roots continually modify the nearby soil structure by their mechanical and metabolic activities. Physical changes induced by roots include, first, compression of the soil at the root surface to create a zone of minimal voids. X-ray diffraction analysis has showed that mineral particles and the pore space between them are smaller in the rhizosphere than in the bulk soil. Second, root surfaces have a high negative water potential caused by the evaporative demand of the aerial parts of the plant. Rhizosphere microorganisms may therefore experience considerable water stress during the parts of the day when evaporative demand by the plant is high. This is particularly true when the hydraulic conductivity of the soil is low or when the roots are widely spaced.

The activities of roots also bring about chemical changes in the nearby soil. First, salts are carried by mass flow to the root surface, and those not absorbed by the plant may accumulate and even precipitate out at the root surface. Rhizosphere microorganisms may therefore be immersed in a medium of low osmotic potential. Second, the pH in the rhizosphere may differ by more than two pH units from that of the bulk soil. When taking up anions such as nitrate, the roots secrete bicarbonate, and when adsorbing cations such as ammonium, the roots secrete hydrogen ions; each process leads to shifts in rhizosphere pH. Third, the oxygen and carbon dioxide levels in the soil atmosphere change due to root respiration. Moreover, some roots exude volatile compounds such as ethylene and terpenes that may inhibit or stimulate microbial growth. Conversely, many plants (e.g., rice) that grow in waterlogged soils contain an aerenchyma that conducts

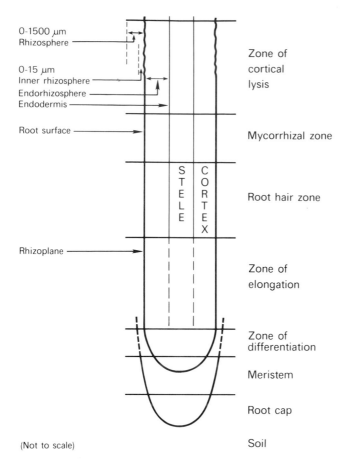

Fig. 1. Major regions of differentiation in the root and rhizosphere. Drawing by G. E. Rinder and R. M. Schuster

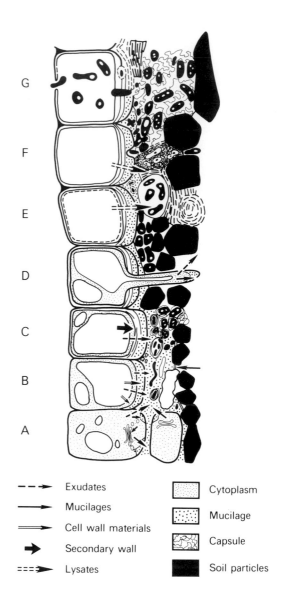

Fig. 2. Stages in the life cycle of a root surface cell and the nearby rhizosphere. At **A,** the root cap zone, there is massive excretion of gel by the dictyosome vesicles into the rhizosphere. At **B,** the epidermal cells are emerging from beneath the root cap and beginning to extend; at this stage, the cell is enclosed only by a primary wall. With the cessation of longitudinal extension, the cell secretes a secondary wall (stage **C**) that the developing root hair must penetrate (stage **D**). Root hairs function for a relatively short time before they and the other epidermal cells autolyse. With mechanical or microbial lysis of the cuticle, further mucilages are released into the soil. These organic materials, together with the products of cell autolysis (lysates) provide substrates for microorganisms, which proliferate in the rhizosphere (stage **E**). These microorganisms break down the primary wall (stage **F**) and invade the epidermis and cortex (stage **G**). At this stage, the outer cortex ceases to function physiologically, though it may persist for weeks and even months. In many annual crops, stage G represents the root and rhizosphere just before harvest and is the final stage in rhizosphere development in the living plant. With the death of the root system after harvest, the root tissues become part of the soil organic matter. Drawing by G. E. Rinder and R. M. Schuster

oxygen from the leaves to the roots, giving a higher redox potential in the rhizoplane than in the bulk soil and causing ferric iron to be precipitated at the root surface.

Changes in the organic matter content of soil at the root surface accompany these physical and chemical changes in the rhizosphere. First, to ease the passage of the root through the soil, the root cap secretes a carbohydrate-rich gel that causes the outermost cap cells to separate and slough off into the soil. Root hairs also secrete mucilage at their tips as they penetrate into fine crevices in the soil. A further source of mucilage is that released by the mechanical rupture of the epidermal cells of the root by soil minerals.

Root exudates provide another source of organic materials in the rhizosphere. Exudates are chemically diverse and include simple low molecular weight compounds such as organic acids, sugars, and amino acids, as well as complex substances such as vitamins and plant hormones. When the cortical cells suberize and autolyse, the products of autolysis (lysates) also enter the rhizosphere. Up to 60% of the net photosynthate of some plants passes down to the root, and more than 30% of this assimilate may be deposited in the soil fabric while the root is still living. In a wheat crop, the amount of carbohydrate deposited in the soil by the roots during their active growth may be greater than the amount deposited in the grain. In many crops, large amounts of root tissue enter the soil as the result of disease, browsing soil animals, or shedding induced by drought, nutrient deficiency, or removal of the aerial shoots by grazing or cutting.

The organic acids and chelating compounds secreted by roots modify nearby soil particles, accelerating the weathering of soil minerals and releasing nutrients in the rhizosphere. Secretion of chelating agents may be increased by particular nutrient deficiencies (e.g., iron and manganese). In certain soils, this acid efflux may release elements such as aluminum in amounts toxic to the root.

These physical and chemical changes induced by root growth and metabolism make the soil near the root very different from the bulk soil. Above all, it is energy rich and supports a great diversity of life and large numbers of microorganisms. Thus, microbial populations reach 1 $\times$ 10^9 cells per cubic centimeter, populations 10–100 times larger than those in the bulk soil. Rhizosphere microorganisms include bacteria, viruses, fungi, arthropods, mites, amoebas, flagellates, yeasts, actinomycetes, and algae. These organisms, together with the root, constitute a specialized ecosystem.

Many rhizosphere microorganisms are not merely passive epiphytes on the growing root; they may secrete materials that modify root morphology and internal anatomy, influence the number of root hairs, and control the type of branching of the root system (mycorrhizas) and the frequency of lateral initiation (proteoid roots). Thus there is constant interplay between the root and the soil microorganisms and between the various types of microorganisms. Ultrastructural and histochemical observations of the root and rhizosphere are summarized diagrammatically in Figures 1 and 2.

The rhizosphere extends to more than 1 mm from the root surface, and because some root pathogens are attracted toward roots 5 mm away, some rhizosphere effects must extend at least this distance from the root. Even within the rhizosphere, there are differences in microbial distribution. Large populations of rhizosphere bacteria can be removed by gently washing roots, but even more bacteria can be obtained if the roots are macerated or shaken with glass beads. These treatments remove microorganisms that are attached to the root or are in the grooves between the root cells and embedded in the gel. Conventional microbiological methods and electron microscopy have shown that, just as the rhizosphere microorganisms differ in kind and number from those in the general soil fabric, the microorganisms at the root surface (rhizoplane) differ from those in the general rhizosphere. Electron microscopy shows that soil bacteria and fungi may occupy the spaces between the cortical cells of healthy roots, and this has given rise to the concept of the "endorhizosphere"—the rhizosphere within the root. In fact, a continuum of microorganisms exists from the endodermis out to the boundary of the rhizosphere.

Despite the huge numbers of bacteria in the rhizosphere, only 7–15% of the actual root surface is occupied by microbial cells. Initially, most of the bacteria lie in the gel contained in the grooves between the host epidermal cells.

The traditional methods of microbiology in which microorganisms are washed from roots and cultured on nutrient agar indicate how many viable bacteria are present that are capable of forming colonies on a particular medium. Some microorganisms are identified by their morphology, staining reactions, and reaction to various biochemical tests. These methods of isolation and characterization, however, give no indication of the location of microorganisms in the soil fabric or of their interactions with each other, with the soil particles, and with plant roots. In contrast, the distribution of the different types of organisms in the soil and at the root surface can be studied by using the electron microscope to examine ultrathin sections of soil; these types of organisms can be distinguished on the basis of morphology and cytology. At present, however, it is not always possible to relate the various cytological types seen by electron microscopy to those obtained by the traditional microbiological methods. Modern histochemical methods may be able to provide information on the biochemistry and function of microorganisms, and in the context of microbial ecology of the rhizosphere, this may be more important than their identity.

In this book, we examine the root-soil interface at magnifications that make visible the fine structure of the microorganisms and the objects in their environment; thus we develop visual concepts of the interrelationships between the soil, root, and microorganisms, and in this way we hope to better understand the ecology of soil microorganisms on the microscale that is the reality of the soil and rhizosphere biology. Soil microorganisms are at the limit of resolution of the light microscope, and they are not visible in or on roots unless they are stained after the roots have been washed free of soil—these factors limit the usefulness of the light microscope for ecological studies.

By combining electron microscopy with the use of undisturbed soil and root samples, either as ultrathin sections for transmission electron microscopy (TEM) or as frozen, dehydrated specimens for scanning electron microscopy (SEM), we can develop a new concept of the ecology of microorganisms at the root-soil interface.

Part I.

The Soil, Root, and Rhizosphere

The Soil

Minerals

Soil remote from the root is dominated by minerals that cause severe technical difficulties in preparation of ultrathin sections of soil fabrics. Quartz grains usually shatter, leaving small fragments that are about micron size and display characteristic conchoidal fracture surfaces. Sometimes the quartz is enclosed in a cutan (skin) of clay minerals that persists in the section because it is penetrated by the embedding medium. During cutting, quartz particles may be pulled out of the soil fabric, leaving angular electron-transparent holes in the plastic embedding medium. In SEM, quartz grains are recognized by their characteristic hexagonal morphology.

The appearance of clay minerals in ultrathin sections depends on the angle at which they are sectioned. Micas cut at right angles to the leaflets show a characteristic booklike appearance. When cut parallel to the crystal faces, they appear as irregular plates with angular outlines. As the clay is weathered, it is reduced to small spindle-shaped particles about $0.05-0.1$ μm long and about 0.01 μm thick at the widest part. These finer clay particles move in the soil, frequently coating other minerals, forming a lining to voids, and enclosing organic matter particles. The platelike particles lie with their flat faces parallel to the surface they enclose and may be six to 10 layers thick. Their orientation means that they can be detected under polarized light, and they are classified as cutans in soil micromorphology. Cutans may protect organic matter from microbial attack (physically protected organic matter) and thus create inaccessible sites from which the organic matter may only be released by tillage or by comminution by soil animals.

Voids

In a typical ultrathin section (100 nm thick) of a soil, the whole fabric is crossed by voids and appears to be much more open than mechanical analysis or light microscopy would suggest. Voids are apparent in ultrathin sections because the sections are much thinner than sections examined with the light microscope where such fine voids are obscured by surrounding materials.

The largest observable voids are several microns in diameter, but because of the high resolution of electron microscopes, voids as small as 0.01 μm across can be detected.

The larger voids include old root channels and the spaces between large mineral grains, and these are randomly oriented. Voids near the root can be narrow and planar and lie tangentially to the root surface. In field soils the finer voids would normally be filled with water; those of submicron size may contain organic matter recognizable by its irregular form and reaction with specific histochemical reagents.

Organic Matter

Detection of organic matter by TEM after conventional fixation depends on the reaction of heavy metals (osmium, lead, and uranium) with a limited number of organic groups unless it is impregnated with metals from the soil. The organic groups that react with heavy metal stains include dihydroxy and trihydroxy phenolic groups, unsaturated alcoholic, aldehydic, and acidic groups, and phenolic syringyl groups derived from lignins, as well as sulfhydryl and aromatic groups on amino acids. Carbohydrates are unstained by conventional TEM techniques and therefore remain undetected except as electron-transparent spaces in otherwise dense mineral fabrics. However, use of histochemical stains specific for neutral and substituted carbohydrates shows that many fine voids contain amorphous polysaccharides that stabilize the soil fabric. Many carbohydrate fragments occur in crevices smaller than the diameter of a soil bacterium; this may explain how biodegradable organic compounds occur in soils but are not metabolized. Such physically protected organic matter in these inaccessible sites may be attacked by microorganisms following cultivation or ingestion by soil animals.

Organic matter in soil can range from histons of newly shed plant and animal tissue in which the cellular structure is hardly changed from that seen in the living state, down to globular and amorphous humic materials of macromolecular size (5–10 nm). Cell wall remnants are distinguished by their characteristic lamellar structure: parts rich in carbohydrate remain electron-transparent after normal fixation, whereas other wall layers such as cuticles, middle lamellae, secondary wall layers, and tertiary lamellae are electron-dense due to the reaction between osmium and the polyphenols they contain. As decomposition proceeds and carbohydrates are removed, more and more phenolic groups are exposed and the material becomes more electron-dense. Middle lamellar material is amorphous, but lignin residues from secondary wall layers may persist as fibrous skeletons. In some soils, conventional osmium staining indicates that small crumbs are held together by such amorphous and fibrous electron-dense materials. In some cases the fibers may only be 10 nm in diameter.

Decomposition of plant material in waterlogged soils is slow because such conditions are not conducive to the growth of decomposer microorganisms. Large amounts of cell wall remnants accumulate as multilamellate convoluted layers intermixed with fine layers of clay and small colonies of bacteria at favorable microsites.

Insects and mites inhabit the upper layers of soils, and their exoskeletal remains may be recognized either by the characteristic morphology of the remnants or by the ultrastructure of the cuticle. Fecal deposits from these soil animals stain densely, often containing recognizable undigested plant cell material, and are associated with bacteria.

Soil bacteria are associated with larger masses of organic matter and form small colonies on and within it, but unlike the bacteria in the rhizosphere, bacteria associated with the soil organic matter are usually devoid of storage materials such as polyphosphate and polyhydroxybutyrate; this indicates that they are growing under low nutrient conditions.

Humic materials are made up of undecomposed organic matter together with certain microbial by-products. Humic materials have been separated into various classes on the basis of their solubilities in acids, alkalis, and organic solvents (humic acids, fulvic acid, etc.). Such chemical distinctions are not possible in ultrathin sections of soils; instead, humic materials are classified by their morphology or their reaction to various histochemical tests. Humic materials are often detected as amorphous masses in the interstices between clay minerals, forming networks between other constituents of soil crumbs and associated decomposing tissues. Their preservation and detection after conventional fixation depend on their reaction with organic aldehydes, which cross-link them in situ, and with osmium and lead, which make them electron-dense and therefore detectable. Sometimes, however, histochemistry and electron probe microanalysis are necessary to distinguish them from amorphous mineral deposits. These distinctions are not possible in most soils because organic and inorganic materials enter into complex combinations. Because humic materials contain a variety of active groups (amino acids, polyphenols, and carbohydrates), stages in the formation of humic compounds from plant residues can be followed by histochemical techniques and TEM. The smallest detectable deposits can be of micromolecular sizes (5–10 nm), so in this way electron microscopy provides a link between the soil micromorphologist and the soil chemist. For example, electron microscopy shows that polyphenols in humic materials often complex with metals and in this way form stable, chemically protected organic matter resistant to microbial enzymes.

Microorganisms

Microorganisms in the soil fabric are mainly in or on organic matter as small colonies or as individual cells completely surrounded by clay fabric. Because these cells are devoid of storage materials, we propose that many of these cells may not be capable of reproduction and growth, although their cytoplasm appears intact. When such cells lyse, however, they provide nutrients for nearby organisms. It is possible that some enzymes from the dead microorganisms are still active. "Phantom" colonies consisting only of the capsule materials that previously surrounded the bacterial cells occur in the outer rhizosphere; the fibrillar or granular carbohydrates derived from these colonies may play an important role in stabilizing the soil microfabric and in the ion exchange properties of the soil.

The Root

Root Cap

The root cap is found on the roots of all higher plants (ferns, gymnosperms, and angiosperms) and consists of a conical mass of tissue that encloses and protects the apical meristem. Cap cells arise by division of cells in an internal meristem that overlies the root meristem. These root cap cells are rich in starch grains, which appear to serve two purposes. First, they detect the direction of the earth's gravitational field and so direct the root to grow downward. Second, the starch acts as a nutrient source for the production of a slimy gel that lubricates the root tip as it forces its way into the soil. Work using ^{14}C-labeled fucose (which is specifically metabolized to form root cap mucilage) has shown that the gel is produced by the swollen vesicles of dictyosomes, which are particularly numerous and active in cap cells. The vesicles move to the plasmalemma and push the carbohydrate through the cell wall onto the root surface and into the intercellular spaces. The gel is rich in substituted carbohydrates, which stain with ruthenium, lanthanum, and manganese, so that it can readily be demonstrated by electron microscopy. Continual secretion followed by hydration of gel forces the cap cells apart so that eventually these cells slough off into the surrounding soil. The cells are filled with starch, which helps them survive in soil for as long as three weeks, during which time they continue to secrete mucilage. The root cap cells then autolyse and collapse, become irregular in shape, and are colonized by bacteria, most of which originates from fragments of organic matter contacted by the root as it grows through the soil.

Unlike the rest of the root surface, the root cap as seen in SEM is generally quite devoid of microbial colonies. This may arise because the cap cells are continuously renewed from within so that new surfaces are continually being exposed and any bacterial cells that have time to multiply are sloughed off into the soil with the cap cells.

Young Epidermis

As the root cap cells are sloughed off and the differentiating tissues behind the root apex enter the phase of rapid longitudinal elongation, the epidermal

6

cells emerge from beneath the cap. SEM shows that at first the surface of the cells is clean, turgid, close-fitting, and devoid of bacterial colonies and embedded soil particles. TEM of ultrathin sections of epidermal cells shows that at this stage the primary wall of the cell consists of two layers. The bulk of the wall consists of mucilaginous pectins and hemicelluloses supported toward the plasmalemma by cellulose microfibrils. The gel is bounded externally by a fine, electron-dense, three-layered membrane called the cuticle.

Although the presence of thick gel layers on roots of certain plants (Ericaceae, apples, corn) has been known for more than 100 years, investigations of the formation and subsequent degradation of the gel have been inhibited by technical difficulties. First, the gel layer is less than 1 μm thick on most of the epidermal cell surface except close to the root cap, in the intercellular spaces of very young roots, and on roots grown in wet soils. Second, mucilage is neither preserved by the fixatives (organic aldehydes) nor stained by the heavy metal salts (osmium tetroxide, uranyl acetate, lead citrate) used to prepare plant tissues for electron microscopy. Unless the gel and the soil fabric embedded in it are fixed before the roots are isolated from the soils, all but the highly polymerized gel in the intercellular spaces and close to the root surface is lost even by gentle washing. To preserve the gel, fixatives must be used that contain agents that cross-link and thus stabilize the gel (e.g., metals such as calcium, magnesium, and lanthanum and dyes such as ruthenium red and alcian blue). Even when such fixatives are used, much of the gel may be lost during preparation for electron microscopy, and it may be necessary to embed the rhizosphere in agar, gelatin, or silica gel before fixation. Third, when sections are dehydrated during staining or preparation of permanent mounts for light microscopy, the gel shrinks until it is beyond the limits of easy resolution. Shrinkage occurs when tissues are prepared for electron microscopy, even with critical point drying, unless the gel has been stabilized. Neglect of these precautions in past work has meant that the function and even the presence of the gel have been overlooked by plant anatomists and physiologists.

When roots are fixed and then isolated from the soil by gentle washing, the surface of the root cap and young epidermal cells is quite clean and smooth. Where bacteria are attached, there is no sign of lysis troughs surrounding the organisms, and even clay minerals do not stick strongly to the surface. In ultrathin sections at this stage, the cuticle is intact, and the root surface is smooth except for remnants of root cap gel and a few adhering clay platelets.

The gel on roots grown in nutrient solutions, on agar, or under wet greenhouse conditions has a uniform texture of granules or fine fibrils or both. By contrast, in field-grown plants the gel may be multilamellate, with layers that stain densely with heavy metals due to changes in texture or chemical composition. A particularly tenacious gel is produced at the root surface when severe desiccation induces cell damage.

Elongation of the epidermal cells causes rearrangement of the microfibrils in the primary wall. From an initial transverse or random orientation, they are passively reoriented until, in the root hair zone, the microfibrils are axially oriented. When the epidermal cells cease elongating, they lay down an inner multilamellate secondary wall in which the microfibrils are in alternate S and Z spirals, and the wall shows a characteristic herringbone pattern in section. The secondary wall may become lignified and thus stain strongly after conventional fixation for electron microscopy. In monocotyledons there is frequently an unequal division of the epidermal cells to produce small, specialized trichoblast cells, which give rise to the root hairs.

Root Hairs

Hairs are produced in strict acropetal succession and frequently occur in ranks. They arise from within the secondary wall and penetrate the primary wall mucilages to emerge into the rhizosphere. Like the epidermal cells that produce them, root hairs have a mucilaginous primary wall enclosed in a fine cuticle: close behind the apex a secondary wall of oriented cellulose microfibrils is produced.

Lectins involved in the host recognition processes (e.g., the legume-rhizobium symbiosis) are secreted at the root hair surface. Infection of the root by nitrogen-fixing actinomycetes (e.g., in *Casuarina, Alnus, Ceanothus*) and vesicular-arbuscular mycorrhizas is also often via the root hairs.

When nutrients are deficient, the root may respond both physiologically and anatomically. In soils where both iron and manganese are deficient, roots may secrete organic acids such as citrate and polyphenols that release ions from soil minerals, chelate them, and aid their transport to the root surface.

In several species (sunflower, green peppers, chickpeas), iron deficiency induces some or all of the epidermal and subepidermal cells to differentiate as transfer cells. In roots, transfer cells are normally restricted to the stele, but when crops are grown in soils of high pH or in nutrient solutions deficient in available iron, fingerlike processes characteristic of transfer cells grow out from the inner face of the tangential walls of the epidermal cells into the cell lumen. Only the outer tangential walls are affected. The lobes are composed of fibrous materials that markedly increase the area of contact between the free space of the cell wall and the plasmalemma; presumably this increased surface area aids the uptake of ions.

The Rhizosphere

The External Rhizosphere

Invasion of the Mucilage

As the root diameter increases and the apex forces a passage through the soil, mineral particles pierce and tear the cuticle that surrounds the primary cell wall; in this way, some mineral particles become immersed in the mucilage of the primary wall. Rupture of the cuticle allows mucilage to flow out into the rhizosphere and embed soil particles and microbial colonies. SEM shows that the surface of such roots is coated with mineral grains that protrude from the gel; also, clay platelets coat the epidermal cells so that the actual surface of the root can no longer be seen. This intimate mixing of the mineral and biological compounds is confirmed in TEM sections showing quartz grains buried deeply in the gel and clay particles that may be located almost at the secondary wall layers. Further confirmation has been obtained through histochemical tests for neutral and substituted polysaccharides, which show that the mucilage fills the soil voids for many microns from the root surface.

Lysis of the Mucilage

Bacterial lysis of the mucilage may be localized at first because parts of the root surface are protected by clay minerals and also because there may have been no contact with microbial colonies to provide the inoculum for root colonization. Because bacteria produce exoenzymes such as pectinases and hemicellulases, lysis troughs are wider than the organisms that produce them. Eventually the lysis troughs fuse so that the mucilage is removed from the root surface. Microfibrils from the deeper layers of the cell wall are then exposed and penetrate the soil fabric. Replicas of the root surface at this stage show a microfibrillar network with bacteria attached. SEM may show the torn cuticle peeling back to reveal colonies in what was the primary wall. At first the bacteria on roots may appear indistinct in SEM because they are coated with mucilage, but the morphology of the bacteria becomes more distinct as the gel is metabolized. In roots infected by some pathogens such as the take-all fungus, the root surface bacteria proliferate because pathogens make the plant cell membranes leaky and on such lesions the mucilage is destroyed by bacteria.

Development of the Rhizosphere

Because roots release so much organic matter as exudates, lysates, and mucilages (50–100 mg/g of root per day) and because the microbial maintenance coefficient is so low (0.03 mg/g/day), microorganisms in contact with the root and in its immediate vicinity are stimulated into active metabolism and reproduction. Based on these figures, each gram of root can support 36 mg of bacteria (2×10^{10} cells). These populations are not normally reached because all points of the root do not contact microorganisms in the soil. Other soil organisms such as fungi, flagellates, amoebas, and nematodes are attracted to roots by organic compounds, carbon dioxide, and volatile compounds released by the root. Therefore, there is a rapid buildup of microorganisms near the root surface. Typical populations are 10^9 bacteria, 10^7 actinomycetes, 10^6 fungi, 10^3 protozoa, and 10^3 algae per gram of rhizosphere soil.

Although rhizosphere soil may contain 50–100 times as many microorganisms as the bulk soil and although the rhizosphere effect (i.e., increased microbial populations compared with background) can be detected as far as 1–2 mm from the root surface, direct observations and theoretical models suggest that most of the microorganisms occur within 50 μm of the root surface. Electron microscopy of old roots where cortical lysis is complete show that populations within 10 μm of the surface may reach $1.2 \times 10^8/cm^3$. Such populations represent a substantial sink for plant metabolites.

Astronomical though these populations may appear, light microscopy of gently washed and stained roots shows that only 7–15% of the actual root surface is occupied by microbial cells. This agrees with calculations based on the average size of a bacterial cell, which show that a 5–10% cover would result from placing the whole rhizosphere population on the surface of a wheat root.

Statistical analysis of the distribution of bacteria on the root has shown that they are not randomly dispersed over the root surface; they are clumped in small colonies and tend to be located where mucilage is thickest or where exudation rates are high. These sites include the grooves between the epidermal cells, the tips and bases of root hairs, and areas where laterals burst through the overlying cortical cells. Lesions caused by disease organisms (fungal pathogens, nematodes) or by root browsers are also sites of preferential microbial colonization. TEM of rhizospheres show that many bacteria are located within 10 μm of the epidermis and are not attached to the root surface by their capsule materials or embedded in the surface gel. Such bacteria would be lost during washing and staining of roots for light microscopy. This loss may partially account for the discrepancy in the observed microbial cover and that calculated from relative exudation rates and microbial maintenance coefficients.

A wide range of species of bacteria, fungi, and actinomycetes have been isolated from roots. Bacteria include members of the genera *Bacillus, Pseudomonas, Erwinia, Caulobacter, Arthrobacter, Micrococcus, Flavobacterium, Chromobacterium,* and *Hyphomycrobium.* However, numerical taxonomy comparing rhizosphere isolates with type cultures of expected genera indicates that most organisms from rhizospheres cannot be assigned to such genera and that the rhizosphere isolates appear to form one large group or a continuum of types whose properties (as used for numerical taxonomy) fall outside those of the groups obtained with type culture bacteria. Bacteria seen in electron micrographs can rarely be identified even though they can be distinguished by shape, size, wall structure, and internal ultrastructure. This is because bacteria are identified by morphology, staining reactions, and biochemical properties rather than ultrastructural features. Also, many bacteria from soil and rhizosphere are pleomorphic. Further TEM studies of roots colonized by known bacteria may assist in the identification of other root colonizers.

Fungal genera found in the rhizosphere include *Alternaria, Aspergillus, Bipolaris, Ceratobasidium, Chaetomium, Cladosporium, Curvularia, Cylindrocarpon, Embellisia, Fusarium, Gaeumannomyces, Helminthosporium, Mortierella, Mucor, Penicillium, Pythium, Rhizoctonia, Syncephalis,* and *Verticillium.*

Many rhizosphere organisms produce extensive capsular material that is morphologically or biochemically distinct from the root mucilage. Capsule materials become so intimately mixed with the surrounding root mucilage as colonies develop that it soon becomes impossible to distinguish the mucilages physically, morphologically, or biochemically. Jenny and Grossenbacher (1963) called this material the "mucigel."

The small sizes of colonies and the large variety of morphologically distinct bacteria at the rhizoplane indicate that they mainly arise from individual founder organisms that are touched by roots; migration of bacteria to the rhizoplane is probably limited. Microorganisms in the rhizosphere live in an energy-rich environment, and hence they differ from those in the bulk soil in that they contain large amounts of storage materials, such as polyhydroxybutyrate, which forms electron-transparent globules, and polyphosphates, which appear as small electron-dense granules.

It would be surprising if an ecosystem so rich in species, individuals, and stored nutrient reserves were not exploited by predators. Among the bacteria at the rhizoplane can be seen predatory bacteria such as *Bdellovibrio* as well as bacterial viruses (bacteriophages). Rhizosphere bacteria and fungi are also browsed by amoebas, flagellates, Collembola, and mites, and such browsing may stimulate plant growth. Use of increasingly complex model systems of roots, bacteria, fungi, amoebas, and nonpathogenic nematodes shows that these predators greatly increase the cycling of phosphorus and nitrogen in the rhizosphere.

The Endorhizosphere

Colonization of Intercellular Spaces

Electron microscopy has established that normal healthy roots grown in soil harbor bacteria in the intercellular spaces of the outer cortex. These colonies may become large when roots are grown in nutrient solutions.

In most cases the intercellular gel is not being utilized by the bacteria (there are no lysis zones), and the bacteria do not induce defense reactions by the host. This indicates that these bacteria are utilizing the exudates that enter the intercellular spaces rather than breaking down the more complex cell wall materials.

Colonization of the Cortex

Many roots are quite ephemeral, and even in healthy plants, much of the cortex is invaded by saprophytes. This process starts with the colonization of sloughed off root cap cells; the root hairs only live for a few days before they collapse and are destroyed by soil bacteria. The cortex is then colonized, beginning with the invasion of the mucilages of the primary wall of the epidermal cells; later the whole cortex is colonized, and in mature parts of the root, only the tissues of the stele are alive, making the endodermis the functional surface of the root. Roots of perennial plants initiate secondary thickening, and with the development of the periderm, the cortical tissues are shed into the soil.

Although bacteria begin to colonize the intercellular spaces of the cortex immediately behind the apex and mycorrhizal fungi are well established in the cortex behind the root hair zone, bacteria do not begin to occupy the lumen of the epidermal and outer cortical cells until the cells have autolysed and are devoid of cytoplasmic contents. Thus cortical cell lysis is spontaneous and genetically controlled and proceeds independently of microbial attack. Ectotrophic and endotrophic mycorrhizas tend to prolong the lives of epidermal and cortical cells by several weeks.

Production of readily metabolized mucilages, exudates, and lysates ceases with the autolysis of the cortex, and then the root tissue is invaded by microorganisms able to metabolize the more resistant components of the lignified and sometimes suberized secondary cell wall layers. The carbohydrate-rich layers of the cell wall are first invaded so that the bacteria are enclosed in the more resistant middle and terminal lamellae. Removal of the bulk of the cell wall causes the cortical cells to collapse and obliterate the lumena of the cells. This endorhizosphere is occupied by a wide variety of bacteria, many of which become filled with polyhydroxybutyrate and polyphosphate granules.

Three types of organisms inhabit the endorhizosphere: those that benefit the host plant, those that damage the host, and those that live off dead root tissues. Beneficial associations are those in which organisms enter into symbiotic relationships with the host and, while using the energy resources of the root, provide some benefit for the host plant. Some ectotrophic mycorrhizas improve the nutrient uptake by roots and physically protect the root against parasites or adverse soil conditions. The microorganisms may provide hormones that stimulate root growth or modify root morphology to contribute to survival of the host. Mycorrhizal hyphae penetrate soil that is inaccessible to the root and thus effectively increase the root surface area. Nitrogen-fixing bacteria and actinomycetes provide the root with nitrogen and make the plant independent of soil nitrogen. The damaging organisms include parasites that invade the host tissue, live off the nutrient resources of the root, or actively digest the host's tissues and also saprophytes that digest any dead tissues in the host and form toxins that adversely affect the host plant.

Mycorrhizas

Apart from a few families of higher plants (e.g., Brassicaceae and Cucurbitaceae), the roots of all land plants form mycorrhizal associations with soil fungi. In general, the thicker the roots are, the more the hosts depend on their mycorrhizas for adequate nutrition. In healthy mycorrhizas, the fungi do not penetrate the endodermis, so the association is limited to the root cortex.

Ectotrophic mycorrhizas are associations in which the fungal partner forms a sheath of hyphae around the root (the mantle) and is confined to the intercellular spaces of the cortex of the host root. In this association, short lateral roots are induced to branch dichotomously several times and mycorrhizas may be classified on the basis of their color and branching characteristics.

The mantle consists of hyphae embedded in a mucilaginous gel that supports an extensive mycorrhizosphere containing a variety of bacteria. The outer layers of the mantle may be empty, but the inner layers are filled with cytoplasm and stored glycogen granules.

Within the root, the hyphae penetrate between the host cells and proliferate in the intercellular spaces to form a characteristic morphological structure called the Hartig net. Here, phosphorus and other nutrients obtained by the fungus from the soil are transferred to the host, and carbohydrates are transferred from the host to the fungus and stored as glycogen.

Endotrophic mycorrhizal fungi occupy the intercellular spaces of the host cortex, penetrate the cell walls, and enter the lumen of the cell to form systems of branching hyphae called arbuscules, but they do not induce morphological or anatomical changes in the root and do not form a sheath around the root. Although the hyphae proliferate inside the host cell, they are separated from the host cytoplasm by the plasmalemma. The ectotrophic mycorrhizal fungi transfer nutrients, especially phosphate, to the host, obtain energy-rich nutrients from the host, and store the nutrients as lipid droplets.

Mycelial Strands

Both endotrophic and ectotrophic mycorrhizas form mycelial strands or rootlike associations of hyphae (rhizomorphs) that grow out from the surface of the root into the surrounding soil. These strands may be differentiated into an outer rind of dead, thick-walled cells that protect and enclose a core of living cells; these cells may be further differentiated into large empty "vessel hyphae" and living hyphae that transport materials from the soil to the root. Having penetrated the soil for 10–15 cm, the cells separate and enter the soil fabric. Isotope labeling experiments have showed that water and nutrients such as phosphorus and zinc are transported from the soil into roots by mycelial strands.

Root Pathogens and Biological Control

It is beyond the scope of a book on the root-soil interface to deal with root pathogens in detail, but because root pathogens must penetrate the rhizosphere to infect the root, an understanding of the behavior of root pathogens in the bulk soil and rhizosphere as well as an understanding of the interaction between roots, pathogens, and rhizosphere organisms are necessary preludes to biological control of pathogens.

As an example, we will look at the take-all fungus (*Gaeumannomyces graminis* var. *tritici*), which attacks cereals and grasses and is susceptible to biological suppressive factors in soil.

There are two field symptoms of the disease. If the roots of seedlings are subjected to a heavy attack by the fungus, severe seedling blight occurs; this form of the disease, first described in Australia in 1862, led to the common name, take-all. In the second form of the disease, aptly described in South Australia as "hay die," the level of infection is less severe and may occur at tillering or flowering, causing the plants to suffer water stress and premature ripening. In both forms of the disease, the root cortex is relatively unaffected, with the hyphae proliferating mainly in the stele where the vessels and tracheids become blocked with an amorphous material and turn a characteristic dark brown to black. Hyphae also occupy the phloem, thus cutting off food supplies to the root meristem, reducing the development of new roots, and eventually causing root death.

When wheat roots grow near a take-all propagule, hyphae grow toward the root, where the fungus produces two types of hyphae—dark runner hyphae that are thick walled and melanized and grow mainly along the root, often in the intercellular grooves between the cells, and thin hyaline hyphae that grow out from the runner hyphae. These infective hyphae penetrate the root and grow toward the stele, proliferating in the xylem and phloem. Frequently the tracheids and even the large central metaxylem vessel become blocked with hyphae or with the amorphous material that surrounds the hyphae.

With the formation of a lesion in the root, the number of bacteria increases markedly on the rhizoplane. Bacteria multiply in the mucilage beneath the cuticle, where they can be faintly seen in SEM. Eventually bacterial and fungal lysis completely destroys the cuticle and mucilage so that the microbial colonies can be seen in SEM. The lysis of the mucilage is so complete that the microfibrils of the inner cell wall layers are exposed. Root decay is therefore a combined process involving both the fungus and the bacteria; such secondary infection of fungal lesions is characteristic of many fungal diseases of plants.

Inside the stele, the hyphae enzymatically degrade the walls of the xylem, producing narrow hyphae that penetrate through the thickened secondary wall layers of the mature tracheids and vessels so that the fungus spreads laterally from cell to cell. Individual hyphae begin to accumulate lipids and to produce a thick protective cell wall that aids the survival of the fungus in the stele of root fragments.

The cortex of these infected roots often becomes colonized with bacteria, many of which are Gram-negative rods that are believed to play a role in the phenomenon known as take-all decline, in which continuous cropping of wheat or barley makes the soil suppressive to take-all. In suppressive soils, fewer hyphae emerge from propagules, and the hyphae that do emerge are colonized by small nonsporing bacterial rods. The bacteria become attached to the fungal hyphae (often end-on) and may cause holes about 1 μm in diameter in the fungal cell wall. The hyphae collapse, lyse, and eventually may be completely destroyed except for wall remnants linked to the root by remains of mucilage and microbial capsule materials.

Preparation of Specimens

Conventional transmission and scanning electron microscopes impose severe limitations on the types of specimens that can be examined. First, specimens for TEM must be very thin (1/20,000 mm). Second, except where the specimen can be maintained at very low temperatures in the electron beam (e.g., at $-170°$C), specimens must be dehydrated. Third, if organic materials are to be seen at high resolution, they must be stained or coated with heavy metals.

Specimens for TEM

Isolated soil components may be simply examined by depositing a very dilute soil suspension on a carbon-coated electron microscope grid. Organic components such as bacteria may need to be shadowed with a heavy metal or stained with a heavy metal salt such as phosphotungstic acid to make details as fine as flagellae visible.

Ultrathin Sections

In preparation of ultrathin sections of whole rhizospheres, the soil fabric is first embedded in agar, gelatin, or silica gel to hold its various components in place during the procedures that follow. The specimen is then fixed in organic aldehydes (formaldehyde, glutaraldehyde, or acrolein or mixtures of these) buffered to the natural pH of the soil. If necessary, dyes (ruthenium red or alcian blue) are added, followed by compounds of polyvalent metals such as lanthanum, calcium, or magnesium that cross-link unstable or soluble components such as low molecular weight carbohydrates. Excess aldehydes are then washed out of the specimen with buffer.

Soil minerals are naturally electron-dense, but biological tissues composed largely of light elements such as hydrogen, carbon, nitrogen, and oxygen must be treated with heavy metals to make them electron-dense and visible in the electron microscope. Hence, specimens must be fixed a second time in osmium tetroxide and uranyl acetate before dehydration in alcohol or acetone. They are then embedded in plastic, usually an epoxy resin that supports the soil and root during sectioning.

Clay and biological materials contain considerable amounts of water, and irrespective of the methods used to dehydrate them, specimens will probably shrink when the water is removed. Critical point drying causes least distortion due to shrinkage and has become the standard method for SEM specimens.

Ultrathin sections (50–100 nm) for TEM are cut on an ultramicrotome, and for most soils, a diamond knife is essential. Certain histochemical reactions are performed at this stage (for example, the periodic acid-silver methanamine stain or periodic acid-thiosemicarbazide silver proteinate stain for neutral carbohydrates). The sections are then mounted on copper grids and may be further stained with uranyl acetate and lead citrate.

Specimens for SEM

For SEM, the specimens are plunged into isopentane cooled by liquid nitrogen slush to snap freeze them. The frozen material may be broken or cut to reveal internal structures and then either freeze-dried or critical point-dried and coated with a heavy metal such as gold.

Part II.

Electron Micrographs

Unless otherwise stated, the transmission electron micrographs are of transverse ultrathin sections of roots and rhizospheres fixed in glutaraldehyde and osmium tetroxide and stained with lead citrate and uranyl acetate. The scanning electron micrographs are of whole-mounted, freeze-dried material shadowed with gold.

Most of the root and rhizosphere specimens are from clover (*Trifolium subterraneum* L.), chickpea (*Cicer arietinum* L.), paspalum (*Paspalum dilatatum* Poir.), pine (*Pinus radiata* D. Don), rape (*Brassica napus* L.), rice (*Oryza sativa* L.), ryegrass (*Lolium perenne* L.), and wheat (*Triticum aestivum* L.).

The Soil

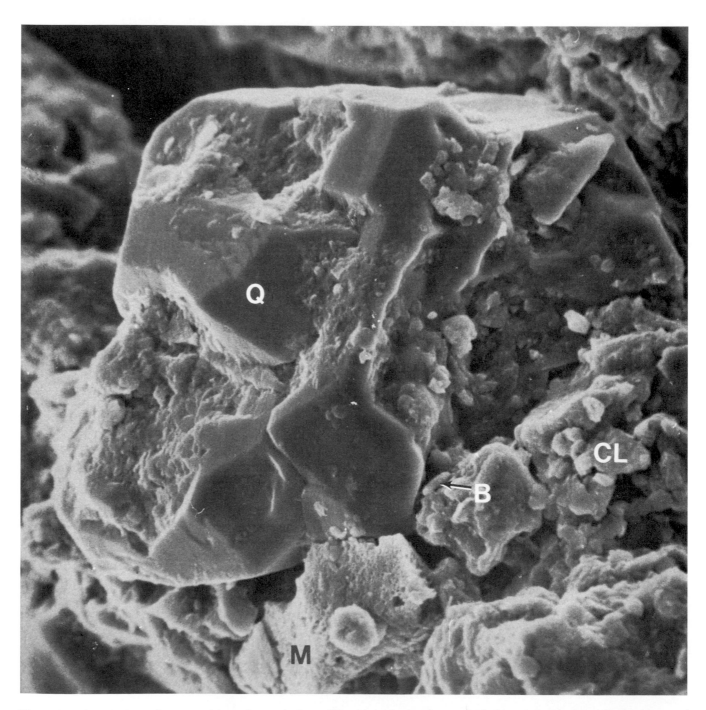

Fig. 3. Sand grain from the outer rhizosphere of wheat. The outer rhizosphere consists mainly of minerals. Quartz sand grains (Q) appear to be both mechanically fractured and chemically etched. The external surfaces are typically smooth and angular. The inner (recessed) surfaces are coated with clay platelets (CL) and amorphous mineral or organic matter. Chemical etching of minerals is enhanced in the rhizosphere by the secretion of acids and chelating agents by roots and associated microorganisms. Bacteria (B) are difficult to recognize in SEM because of capsular materials and coatings of clay particles. M = organic matter, probably root mucilage. (×4,800) Reprinted, by permission, from R. Campbell and A. D. Rovira, The study of the rhizosphere by scanning electron microscopy, Soil Biol. Biochem. 5:747-752, Fig. 10, ©1973, Pergamon Press, Oxford.

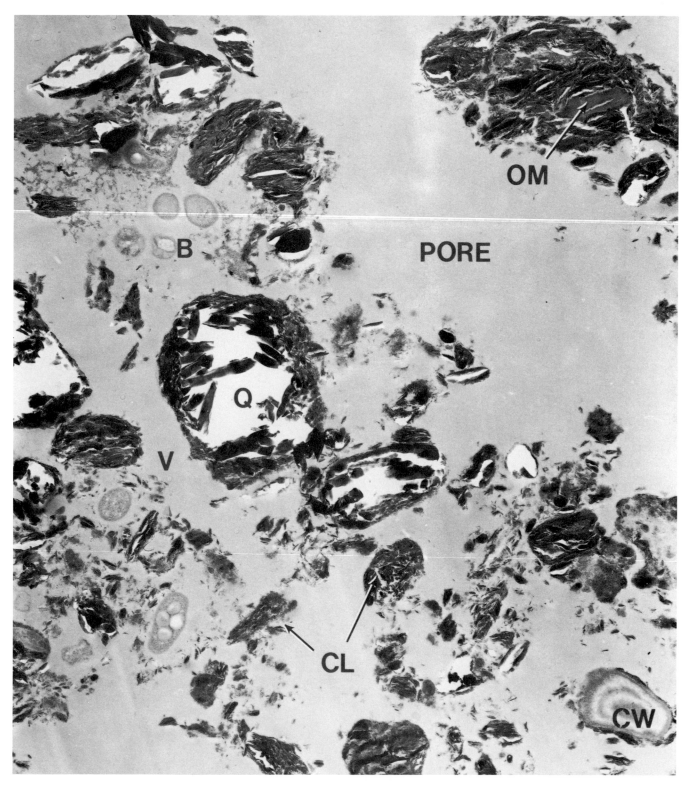

Fig. 4. Outer rhizosphere of wheat. The quartz grains (Q) are frequently shattered into conchoidal fragments during sectioning. Many quartz grains are coated with clay (CL) minerals. Some clay particles form aggregates with other soil components, and such aggregates often have a core of organic matter (OM). B = bacterium, CW = cell wall remnants, V = void. (×16,000)

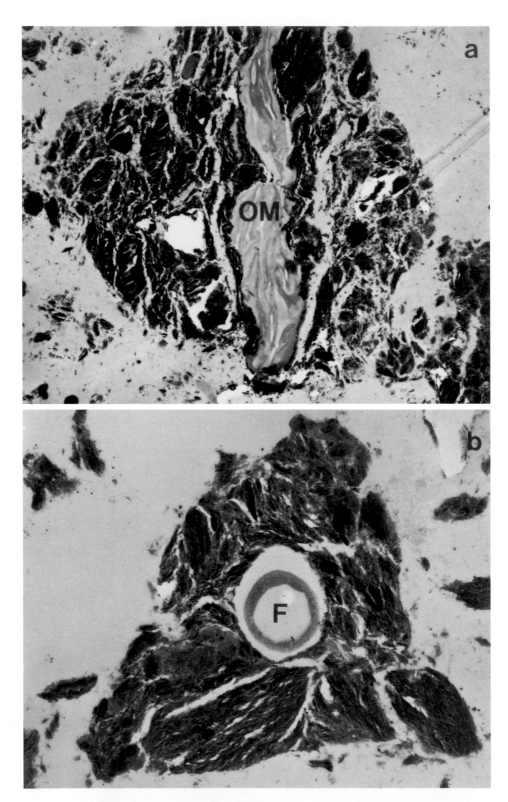

Fig. 5. Physically protected organic matter (OM) in the rhizosphere of **a,** rice and **b,** wheat. Many large irregular grains in soil are composites of clay particles arranged in a random orientation round a central core. In **a,** the organic core consists of plant cell wall materials (×33,000). In **b,** the aggregate is probably held together by carbohydrates secreted by the fungus (F). These organic inclusions are probably physically protected from microbial decay until the aggregate is disrupted by processes such as tillage or comminution by soil animals (×7,200).

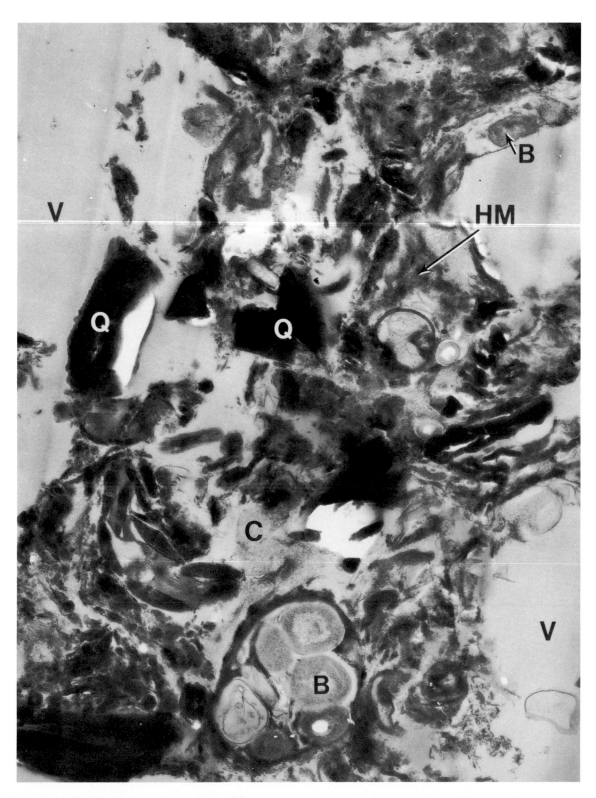

Fig. 6. Soil crumb in *Paspalum* rhizosphere. The various organic and mineral particles in soil crumbs are bound together by humified organic matter (HM) and capsule materials (C) produced by small bacterial colonies (B). The humified organic matter probably constitutes the chemically protected organic matter fraction of the soil. Most crumbs contain small voids (V) that aid aeration of the crumbs. Q = quartz. (Lanthanum stain, ×14,000)

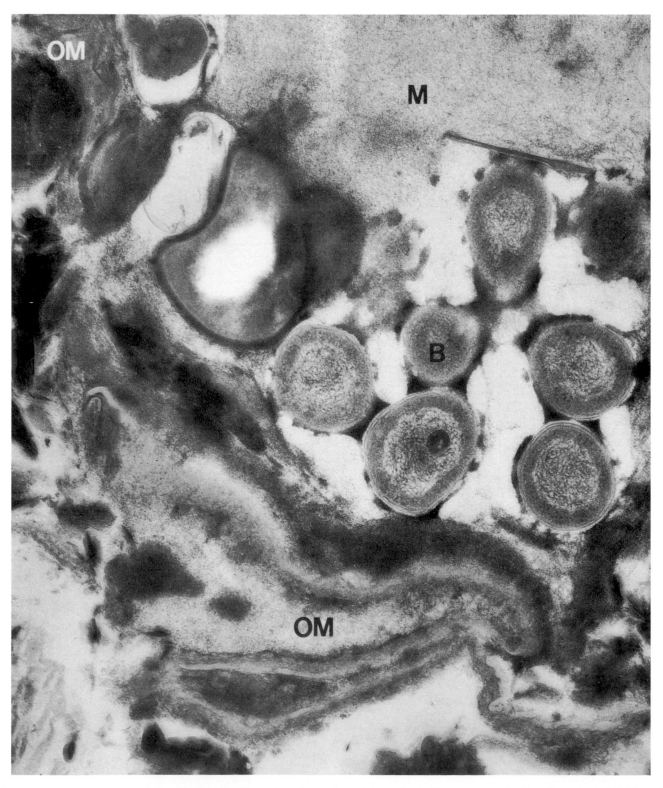

Fig. 7. Bacterial colony associated with mucilage in a *Paspalum* rhizosphere. Fragments of cell wall material (OM) and mucilage (M) from the nearby root support a small colony of bacteria (B). (Lanthanum stain, ×49,000)

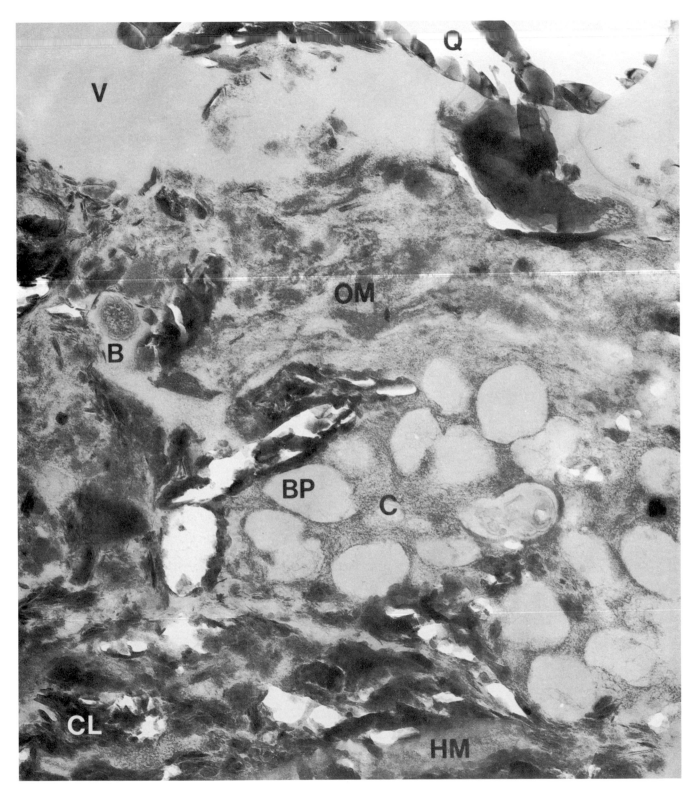

Fig. 8. Phantom colony in *Paspalum* rhizosphere. The microorganisms utilize the available substrate, then die, and autolyse. The cytoplasm and cell walls disappear, but the former location of the colony is indicated by their extracellular capsules (C), which retain the shape of the original (phantom) colony. These capsule remnants hold the nearby soil components—sand grains (Q), clay (CL), and humified organic matter (HM)—together to form a stable soil fabric. B = live bacterium, BP = bacterial phantom, OM = organic matter, V = void. (Lanthanum stain, ×31,000) Reprinted, by permission, from R. C. Foster and J. K. Martin, In situ analysis of soil components of biological origin, Fig. 6, in Soil Biochemistry, vol. 5, E. A. Paul and J. N. Ladd, eds., ©1981, Marcel Dekker, Inc., New York.

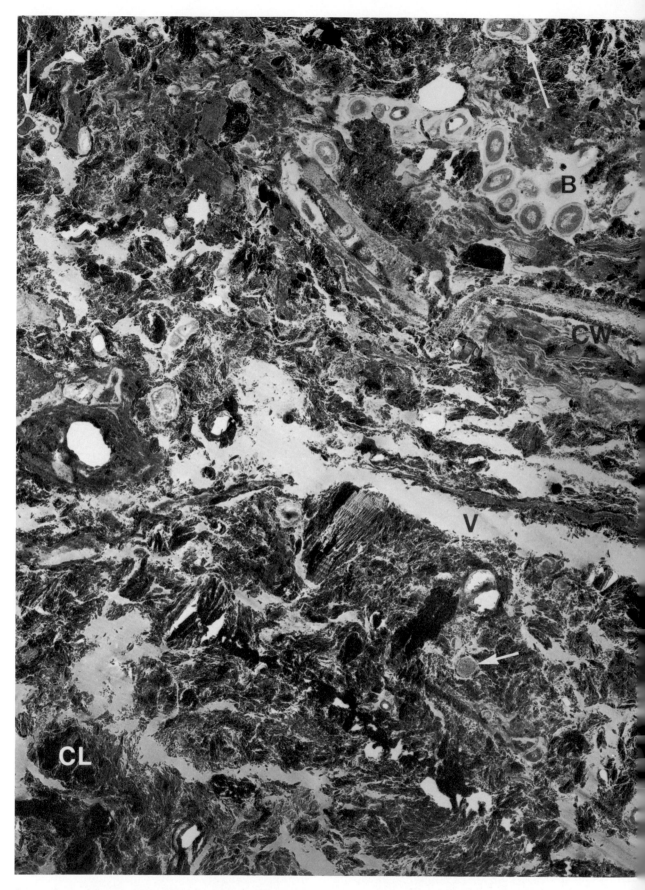

Fig. 9. Clover rhizosphere. A low-power view of an outer rhizosphere shows a soil dominated by clay (CL). Planar voids (V) pass tangentially to the root surface (not shown). Because of root pressure the voids are only 1–5 μm wide. Old cell wall remnants (CW) support a small colony of bacteria (B), and small isolated bacteria are scattered throughout the soil fabric (arrows). HM = humified organic matter, OM = soil organic matter. (Lanthanum stain, ×10,300) Reprinted, by permission, from R. C. Foster, Ultramicromorphology of some South Australian soils, Plate 1, in Modification of Soil Structure, W. W. Emerson, R. D. Bond, and A. R. Dexter, eds., ©1978, John Wiley & Sons, New York.

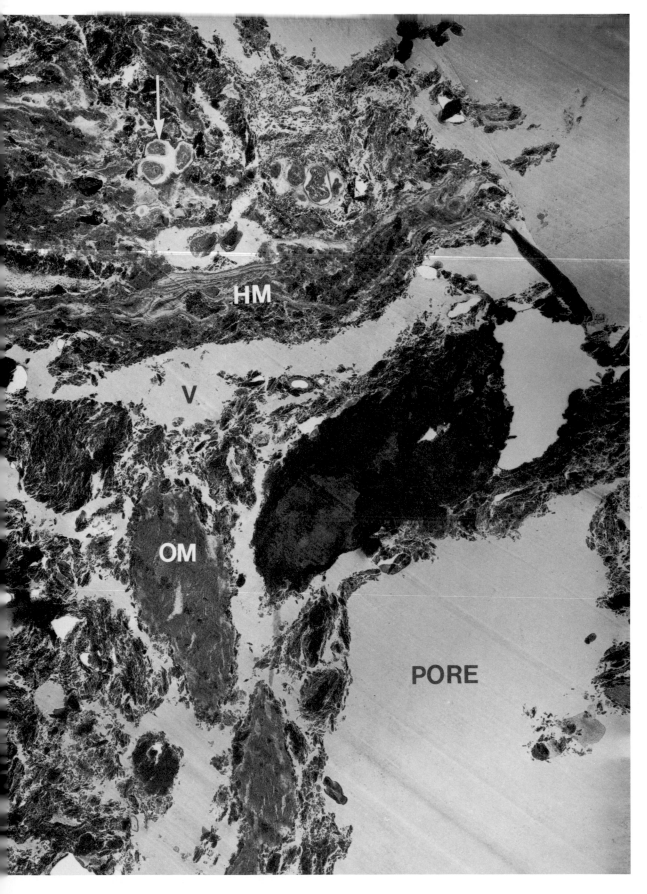

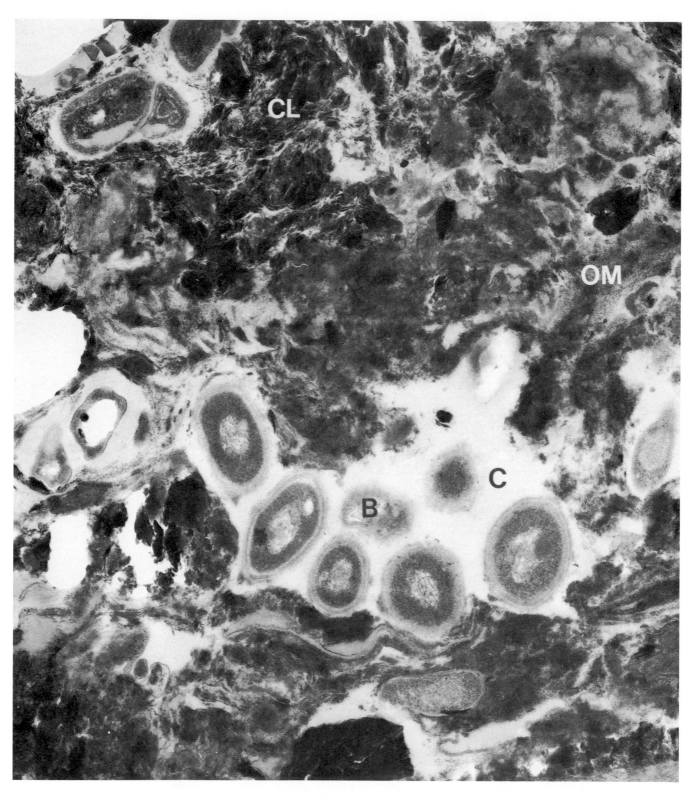

Fig. 10. Clover rhizosphere. Detail of part of Fig. 8 showing the microbial colony associated with the cell wall remnants and amorphous (OM) and lamellate humified organic matter. The capsule material (C) surrounding these bacteria (B) has not stained, but its presence is indicated by the electron-transparent space that the surrounding clay (CL) minerals and humified organic matter have been unable to penetrate. The microbial cytoplasm appears to be unusually dense, with no storage materials, indicating that the substrate may be limiting the growth of bacteria. (Lanthanum stain, ×30,000) Reprinted, by permission, from R. C. Foster, Ultramicromorphology of some South Australian soils, Plate 1, in Modification of Soil Structure, W. W. Emerson, R. D. Bond, and A. R. Dexter, eds., ©1978, John Wiley & Sons, New York.

Fig. 11. Wheat rhizosphere. The complexity of a soil fabric near a root is apparent in SEM. The individual clay particles are mixed with amorphous material that may be mineral or organic in origin. An actinomycete (A) is growing over a clay particle. M = mucigel. (×2,300) Reprinted, by permission, from R. Campbell and A. D. Rovira, The study of the rhizosphere by scanning electron microscopy, Soil Biol. Biochem. 5:747-752, Fig. 8, ©1973, Pergamon Press, Oxford.

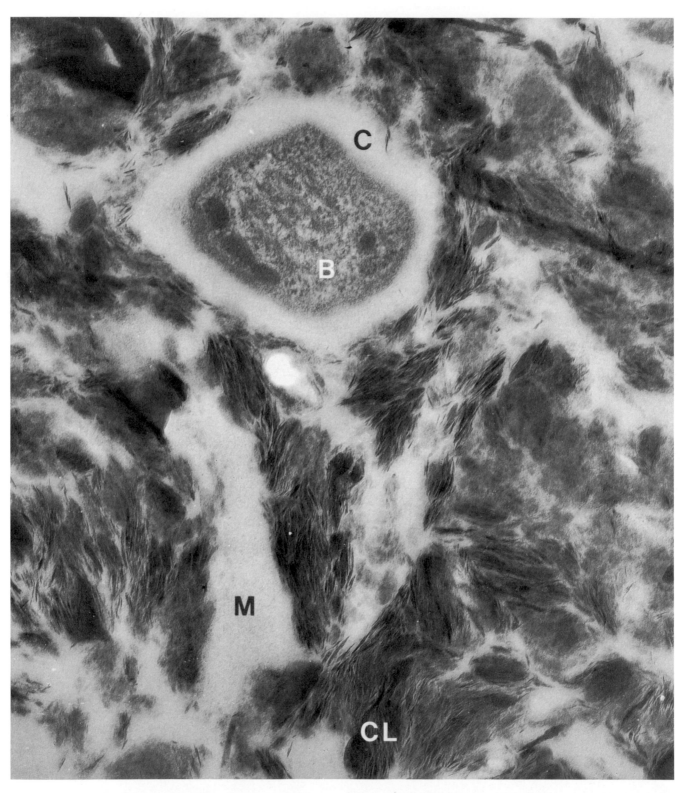

Fig. 12. Detail of clay fabric in a clover rhizosphere. This high-resolution TEM of a clay (CL) fabric contains an isolated bacterium (B) surrounded by unstained capsule material (C). Small voids are filled with mucigel (M). The individual clay particles that make up the aggregates are clearly resolved; the smallest are about 0.1 μm long and 0.01 μm thick and are spindle-shaped in section, tapering toward the edges. (×82,000) Reprinted, by permission, from R. C. Foster, Ultramicromorphology of some South Australian soils, Plate 4, in Modification of Soil Structure, W. W. Emerson, R. D. Bond, and A. R. Dexter, eds., ©1978, John Wiley & Sons, New York.

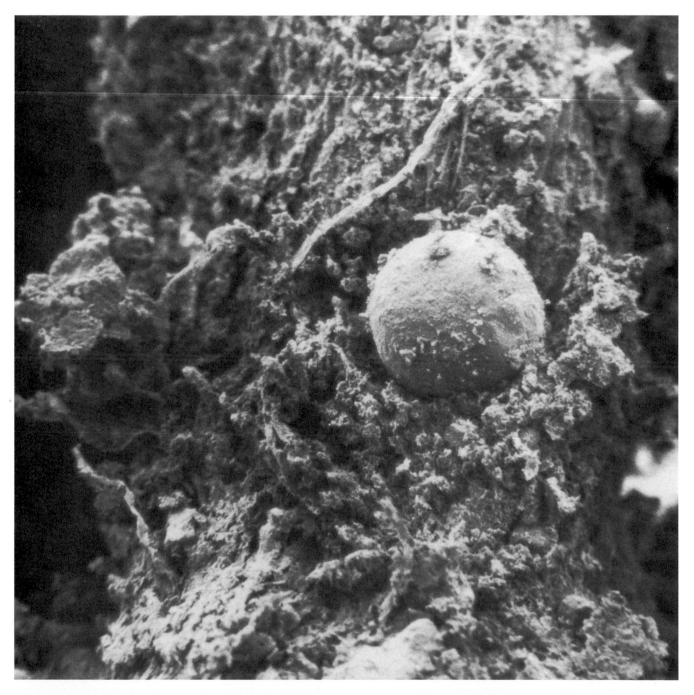

Fig. 13. Fecal pellet on a clover root. Because soils contain large populations of invertebrate animals, fecal pellets are common constituents of soils. This figure shows a fecal pellet adhering to a root surface coated with clay. (×440)

25

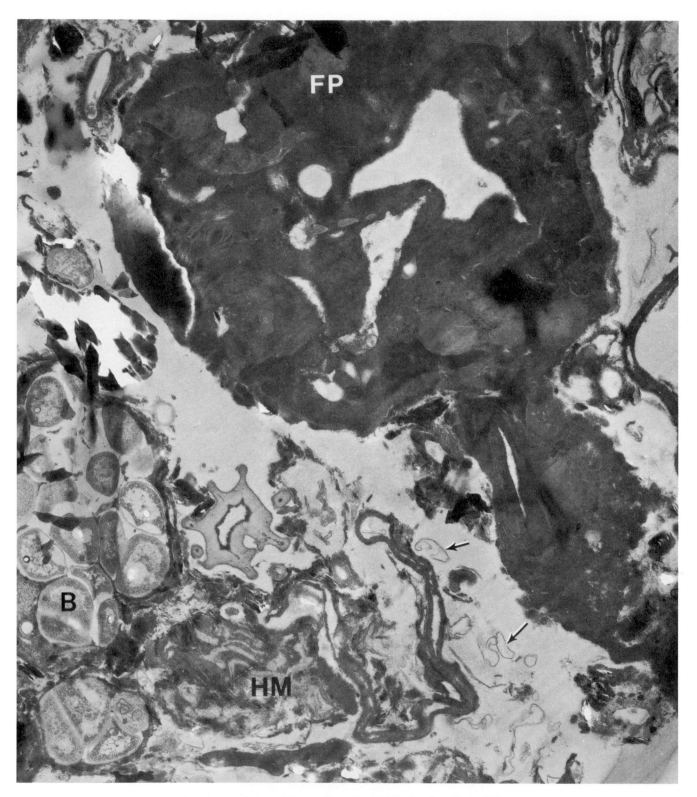

Fig. 14. Old fecal pellet from the rhizosphere of *Paspalum*. In TEM, fecal pellets (FP) stain intensely because of the enzymes they contain. They often contain cellular remnants in which individual cytoplasmic organelles (e.g., chloroplasts) can be observed. Associated with this somewhat decomposed pellet are bacterial colonies (B), remains of plant cell walls, humified organic matter (HM), and fragments of cytoplasmic membranes (arrows). (Lanthanum stain, ×15,000) Reprinted, by permission, from R. C. Foster and J. K. Martin, In situ analysis of soil components of biological origin, Fig. 7, in Soil Biochemistry, vol. 5, E. A. Paul and J. N. Ladd, eds., ©1981, Marcel Dekker, Inc., New York.

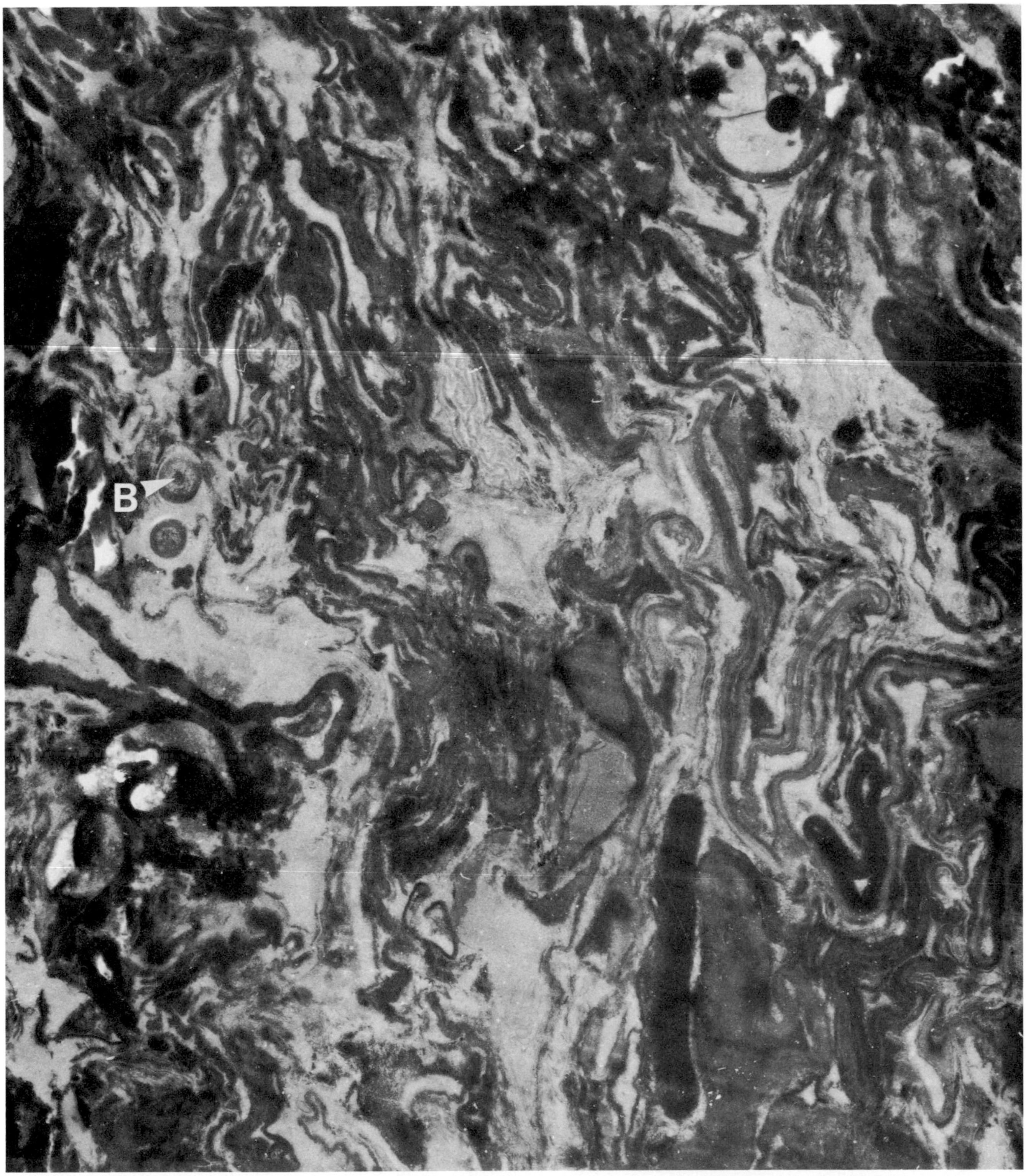

Fig. 15. Outer rhizosphere of ryegrass grown under irrigation. Under wet conditions, lack of oxygen in the soil inhibits microbial growth, and the breakdown of tissues may be retarded. In this case, the soil is composed largely of cell wall fragments in various stages of decay. The cellulose layers are removed first, leaving the polyphenol-rich and therefore darkly staining lignified layers. B = bacterium. (Lanthanum stain, ×15,000) Reprinted, by permission, from R. C. Foster, Ultramicromorphology of some South Australian soils, Plate 2, in Modification of Soil Structure, W. W. Emerson, R. D. Bond, and A. R. Dexter, eds., ©1978, John Wiley & Sons, New York.

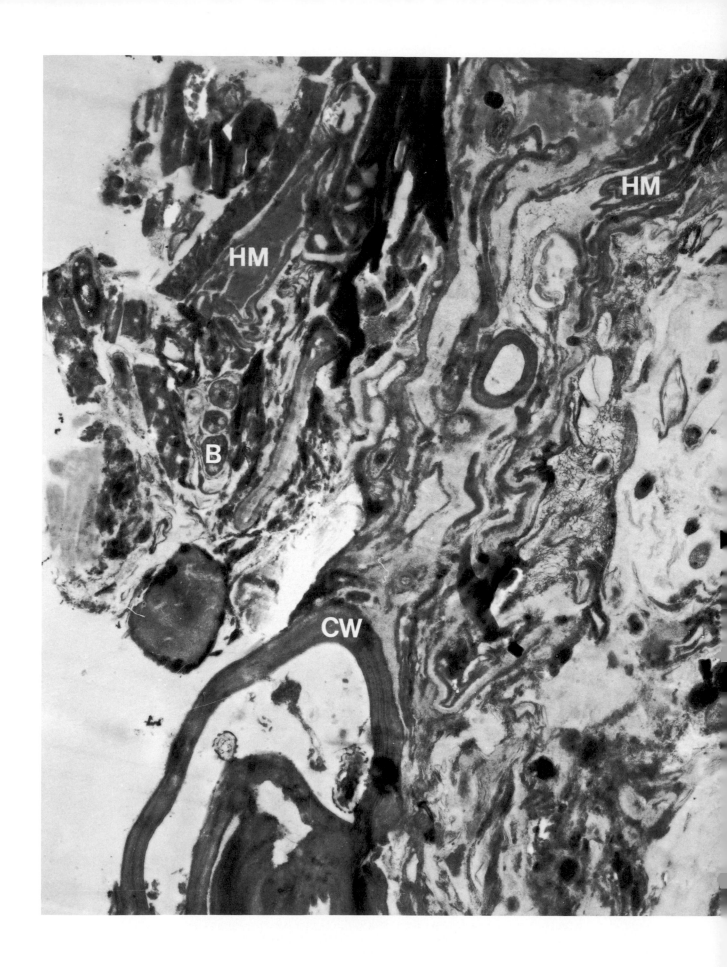

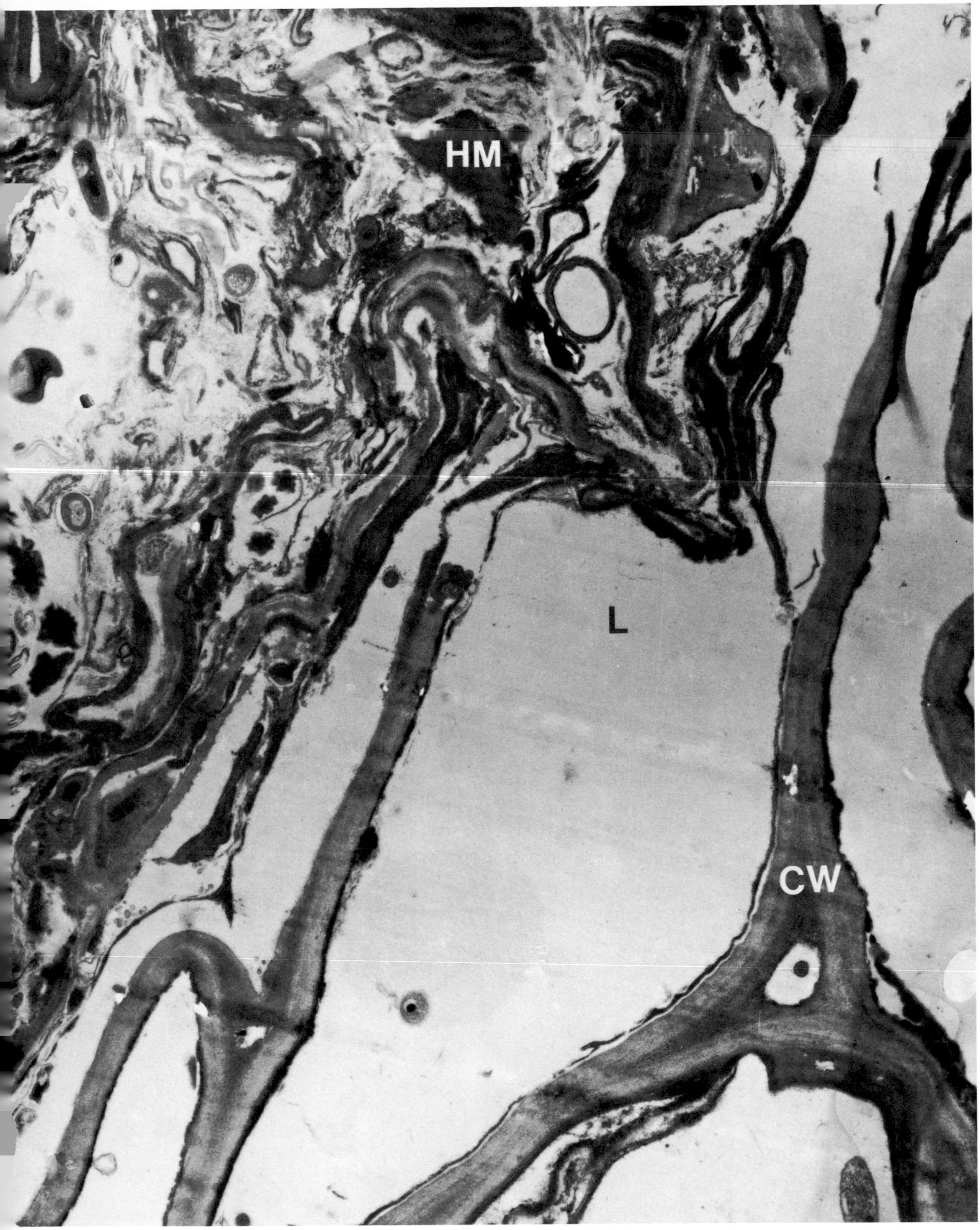

Fig. 16. Inner rhizosphere of ryegrass in irrigated pastures. The bulk of the rhizosphere consists of partially decomposed plant tissues. These range from amorphous electron-dense humic materials (HM) smaller than a micron to recognizable cell wall (CW) fragments several microns long. Electron density is due to the reaction between the osmium and polyphenols derived from lignin residues. Despite all the organic matter, bacteria (B) are relatively sparsely scattered through the fabric, but the usual lobed forms characteristic of the rhizoplane are apparent, surrounded by unstained capsule materials. L = cell lumen. (Lanthanum stain, ×14,200)

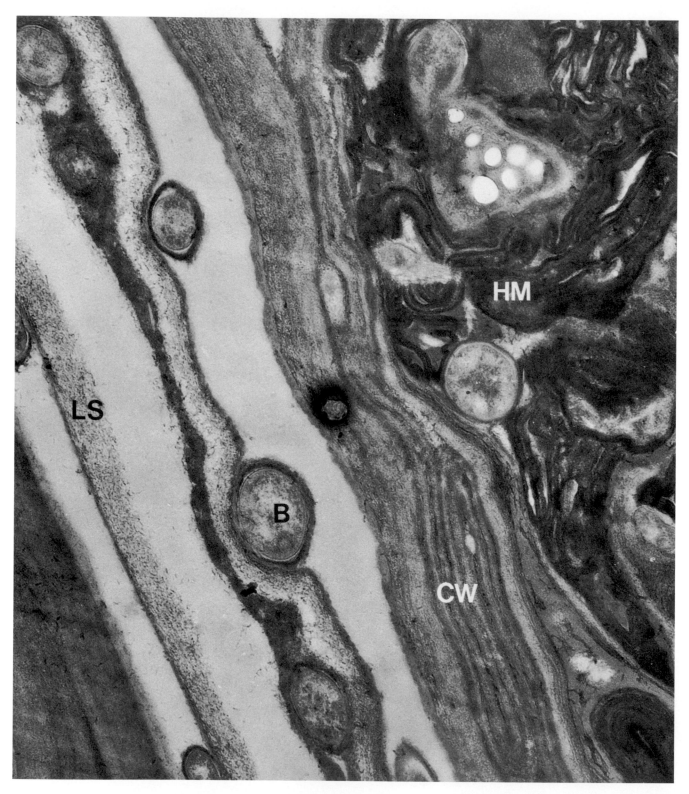

Fig. 17. Old wheat root remnants. The stages in cell wall degradation by soil microorganisms are shown in detail. At the lower left is an almost unmodified double cell wall layer between two highly lignified cells separated by the densely staining middle lamella. Microbial enzymes remove the carbohydrates, leaving the fibrous lignin skeleton (LS). At the right are crushed and folded resistant middle and terminal lamella materials. B = bacterium, CW = cell wall remnants, HM = humified materials. (×79,000) Reprinted, by permission, from R. C. Foster and J. K. Martin, In situ analysis of soil components of biological origin, Fig. 3c, in Soil Biochemistry, vol. 5, E. A. Paul and J. N. Ladd, eds., ©1981, Marcel Dekker, Inc., New York.

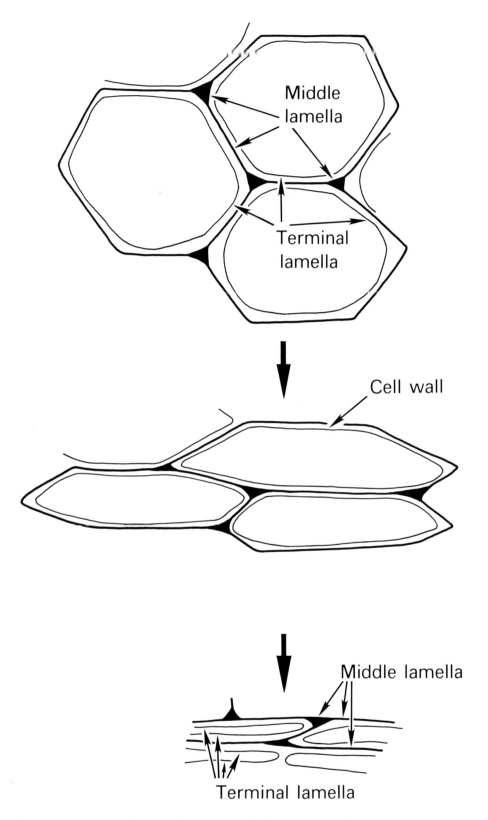

Fig. 18. Stages in degradation and collapse of tissues seen in Fig. 17. After the cytoplasm has lysed, plant cells consist of three layers, cellulosic cell wall, the middle lamella, which glues the cells together, and denatured cytoplasmic remnants, which line the cell lumen and constitute the terminal lamella. The middle and terminal lamellae tend to be impregnated by polyphenolics. This makes them electron-dense after heavy metal staining and also resistant to microbial decay. As microorganisms remove carbohydrate fractions of the cell walls, the cells collapse to give multilamellate complexes largely composed of terminal and middle lamellae. Drawing by G. E. Rinder and R. M. Schuster

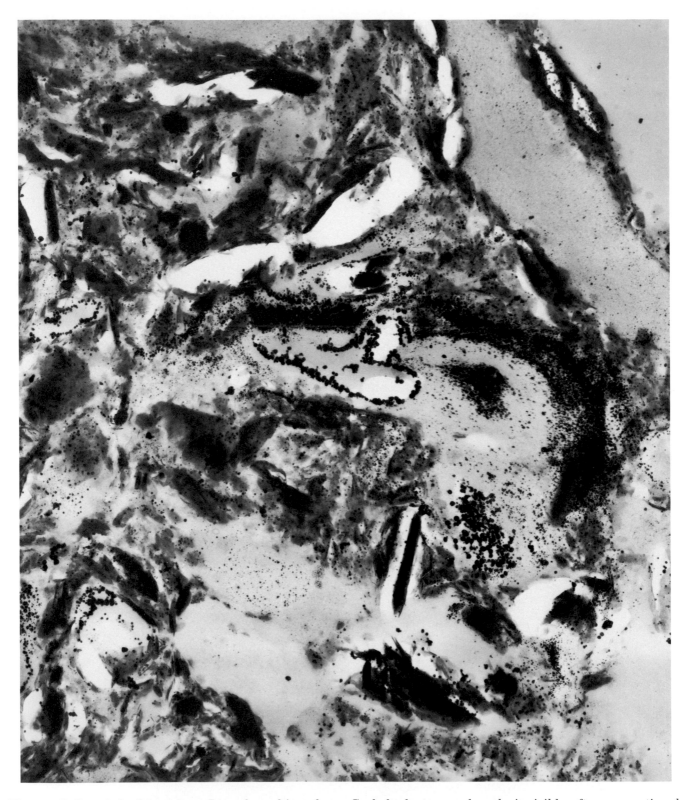

Fig. 19. Soil carbohydrates in a *Paspalum* rhizosphere. Carbohydrates are largely invisible after conventional preparation for TEM, but they can be located as electron-dense (black) deposits of silver by using specific histochemical tests. Here the periodic acid-silver methanamine stain reveals a plant cell wall remnant, cell walls of bacteria, and carbohydrates that cannot be identified with a particular morphological entity. Carbohydrates may coat clay platelets or occur in the fine interstices between clay particles. (×34,000) Reprinted, by permission, from R. C. Foster and J. K. Martin, In situ analysis of soil components of biological origin, Fig. 8b, in Soil Biochemistry, vol. 5, E. A. Paul and J. N. Ladd, eds., ©1981, Marcel Dekker, Inc., New York.

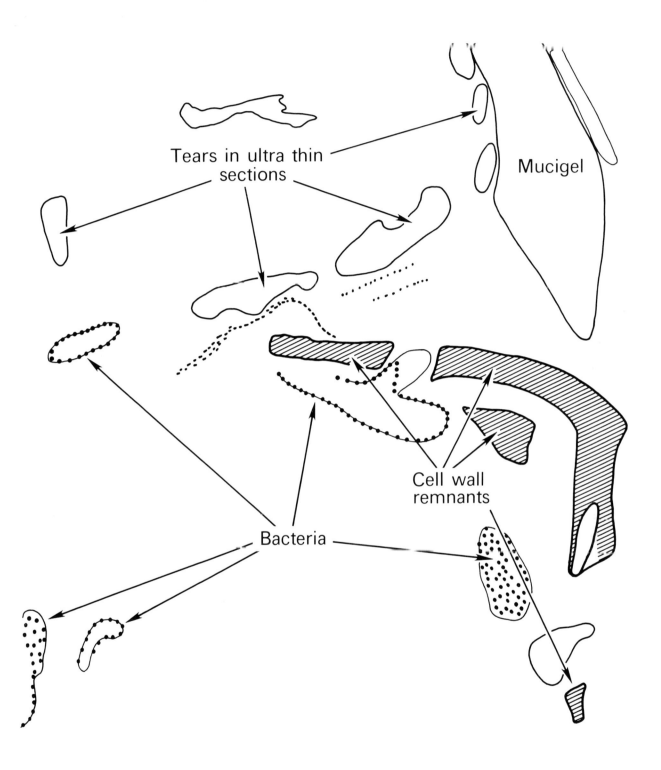

Fig. 20. Diagram of structures in Fig. 19. The nature of the electron-dense reaction product may be typical of the source of carbohydrate. Bacterial cell walls induce the formation of coarse granules; plant cell walls induce a much finer deposit of silver. Drawing by G. E. Rinder and R. M. Schuster

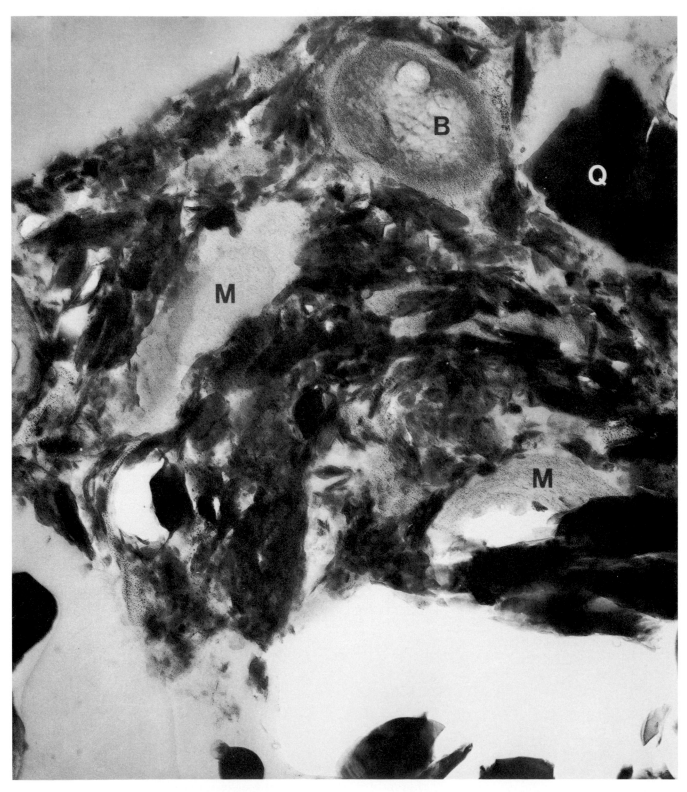

Fig. 21. Soil carbohydrates in a *Paspalum* rhizosphere. Another histochemical test for neutral carbohydrates (periodic acid-thiosemicarbazide-silver proteinate stain) shows that many of the small voids ($< 1\mu$m) between the clay minerals are filled with polysaccharide (M). Some of these voids are too small for bacteria to enter so that the organic materials will be physically protected from decomposition. B = bacterium, Q = quartz. ($\times$70,000) Reprinted, by permission, from R. C. Foster, Polysaccharides in soil fabrics, Science 214:665-667, Fig. 6, ©1981, American Association for the Advancement of Science, Washington, DC.

The Root and Rhizosphere

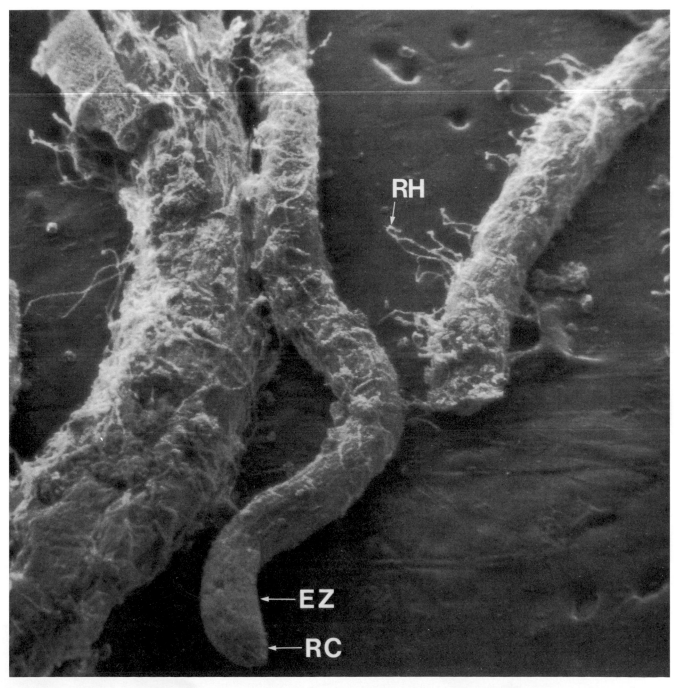

Fig. 22. Wheat roots. A main axis is at the left side of the micrograph. The root cap zone (RC) is followed by the extension zone (EZ). When extension has ceased, root hairs (RH) are formed. (SEM, ×75)

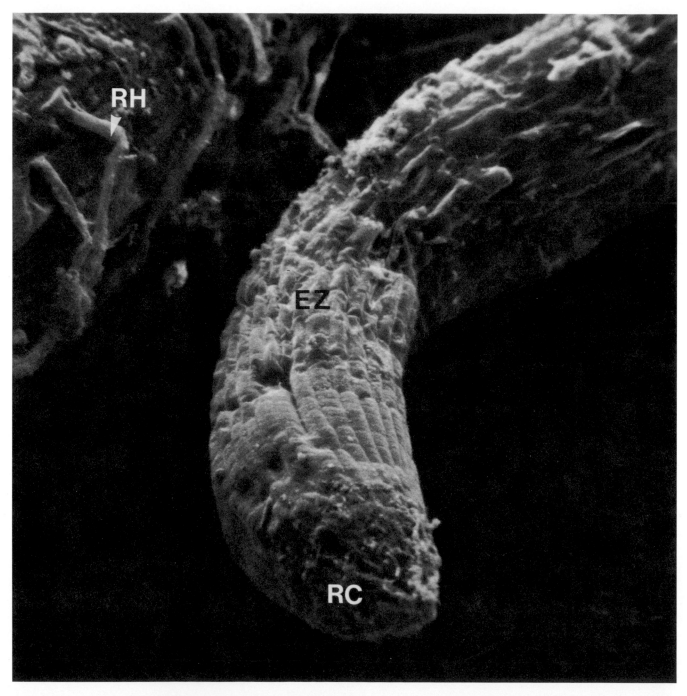

Fig. 23. Root apex of wheat. SEM of the root apex shows the first three zones to be differentiated at the root tip, the root cap (RC) coated with mucilage, the newly divided, bricklike young epidermal cells, and the long narrow epidermal cells in the zone of elongation (EZ). RH = root hair. (×280) Reprinted, by permission, from A. D. Rovira and R. Campbell, Scanning electron microscopy of microorganisms on the roots of wheat, Microb. Ecol. 1:15-23, Fig. 1A, ©1974, Springer-Verlag, New York.

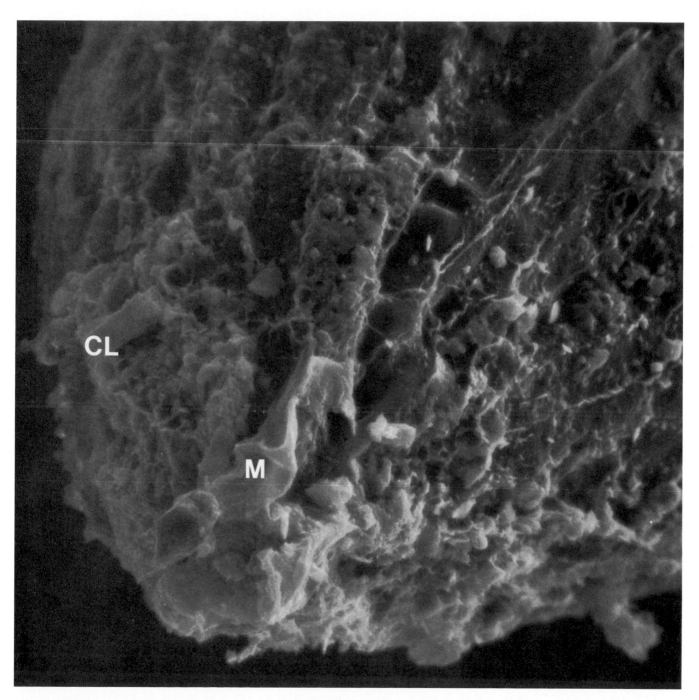

Fig. 24. Wheat root apex. The root cap mucilage (M) is highly hydrated and tends to collapse when the specimen is dehydrated before insertion in the electron microscope. As the root grows forward, it forces its way into the soil fabric and becomes coated with soil particles (CL) that become embedded in the gel. (×1,400) Reprinted, by permission, from A. D. Rovira and R. Campbell, Scanning electron microscopy of microorganisms on the roots of wheat, Microb. Ecol. 1:15-23, Fig. 1B, ©1974, Springer-Verlag, New York.

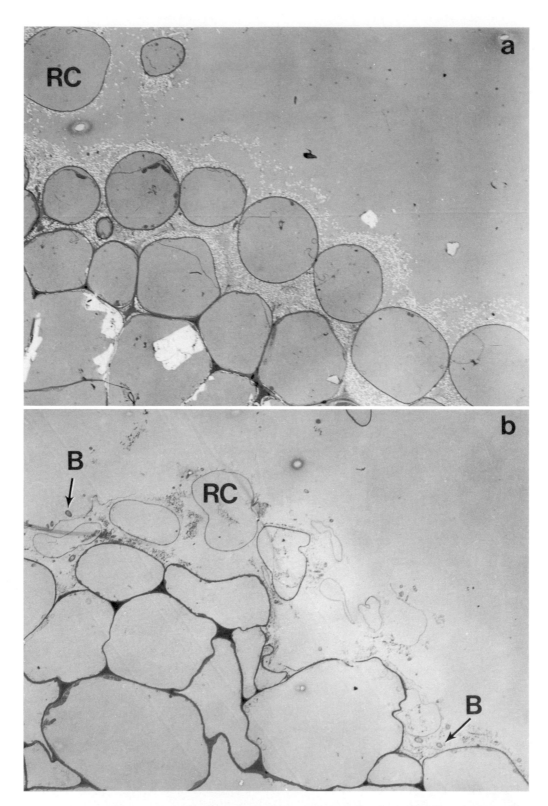

Fig. 25. Chickpea root cap. **a,** Young root cap. The cap cells secrete mucilage on all sides and become separated from each other and from the root to slough off into the soil. Each root cap cell (RC) is surrounded by mucilage. Mucilage is invisible after normal fixation, but its presence can be demonstrated by infiltrating the mucilage with a suitable soluble ion (one of the components of a nutrient solution). This is then precipitated in situ by fixation in organic aldehydes containing the soluble salt of a suitable electron-dense metal. The extent of the precipitate indicates the extent of the mucilage. Here a precipitate of lanthanum phosphate demonstrates the mucilage between the cap cells and shows that the mucilage extends 10–20 μm from the root surface. (×1,400) **b,** Old root cap. Older cap cells (RC) lose turgor, autolyse, and collapse, and bacteria (B) begin to colonize the mucilage. (×1,600)

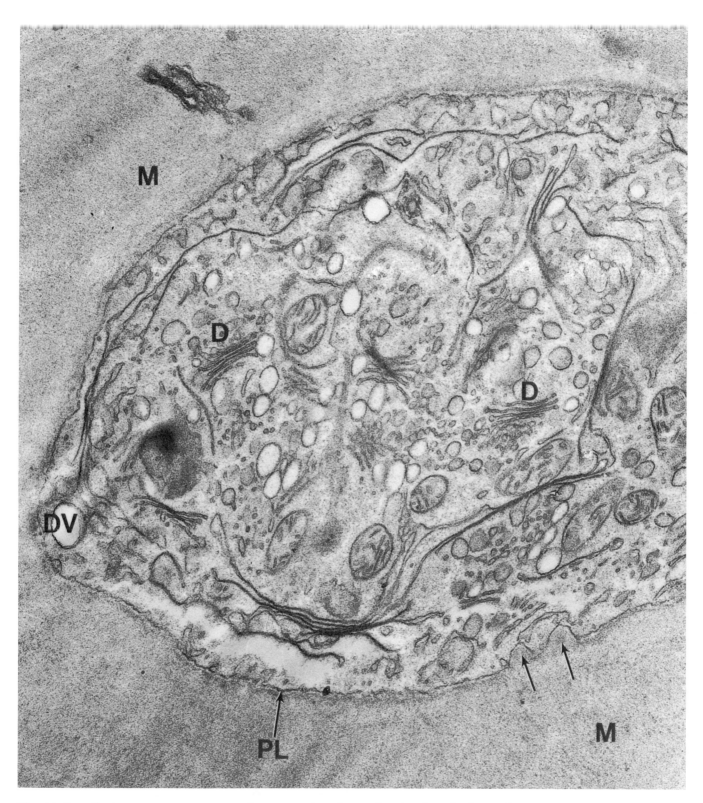

Fig. 26. Detail of a root cap cell of pine, demonstrating the mechanism by which mucilage is secreted. The peripheral cytoplasm of cap cells is filled with dictyosomes (D) that secrete the mucilage (M) into small membrane-bound vesicles (DV). These vesicles move to the surface of the cell (arrows) where the vesicle membrane fuses with the plasmalemma (PL). The membrane breaks down and the mucilage is deposited into the free space of the cell wall (arrows). At first, the contents of the vesicles are electron-transparent (white), but as the vesicles approach the plasmalemma, the mucilage is biochemically changed so that it takes up heavy metal ions and stains grey with potassium permanganate. (×42,000)

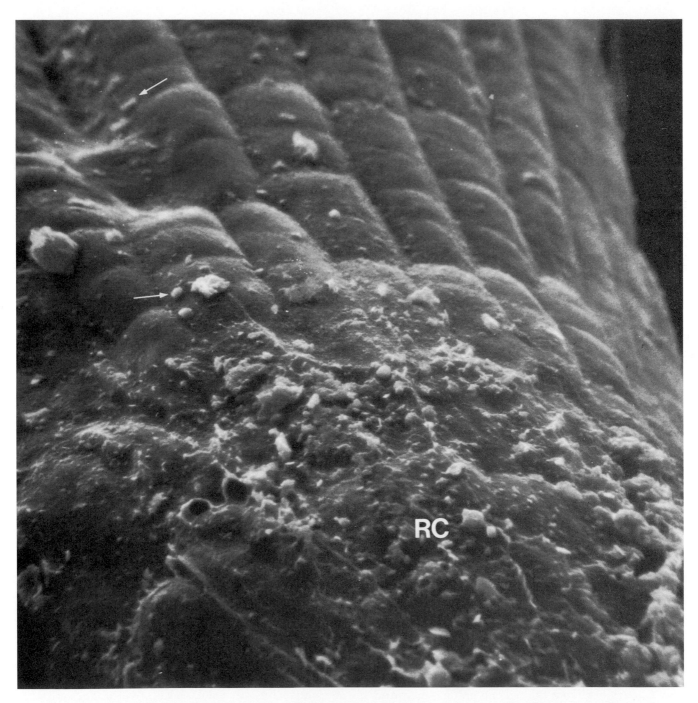

Fig. 27. Detail of boundary between a wheat root cap and the emergent epidermal cells. The root cap (RC) cells are covered with an irregular layer of mucilage, remnants of which also adhere to the young, unextended epidermal cells beyond. A few bacteria have attached themselves to the root surface (arrows). (×1,400) Reprinted, by permission, from A. D. Rovira and R. Campbell, Scanning electron microscopy of microorganisms on the roots of wheat, Microb. Ecol. 1:15-23, Fig. 1A, ©1974, Springer-Verlag, New York.

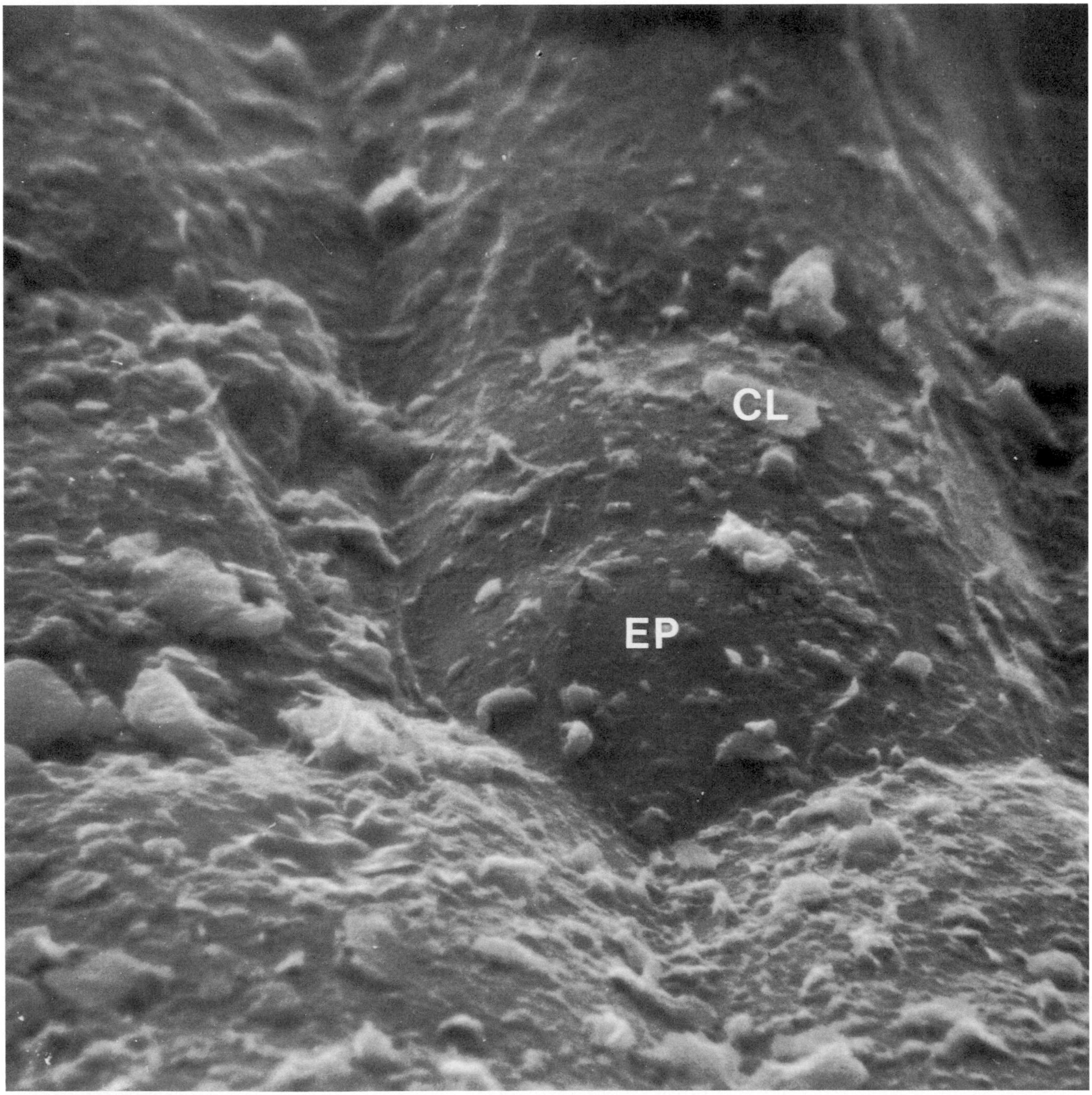

Fig. 28. Detail of boundary between root cap and epidermal cells of a wheat root. The root cap cells in the foreground are covered with mucilage. The emergent epidermal cells (EP) have remnants of mucilage from the cap-epidermis interface. CL = clay mineral. (×6,000)

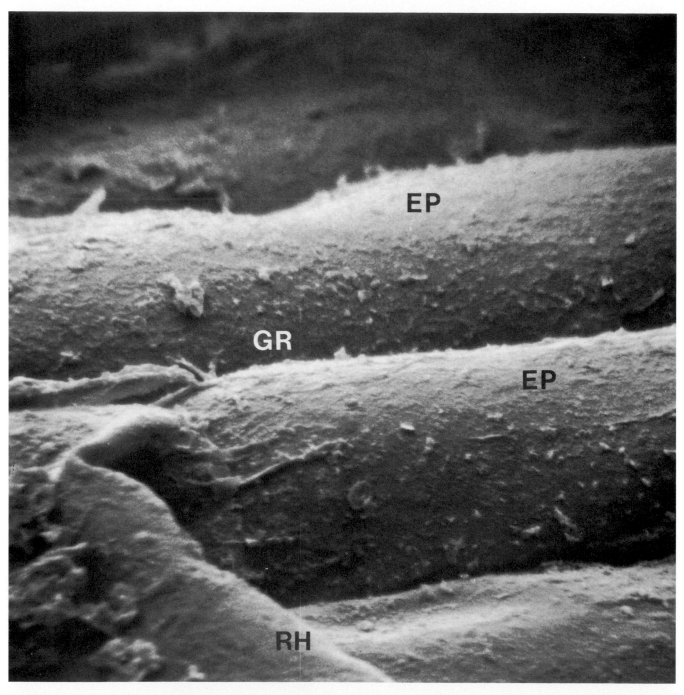

Fig. 29. Fully differentiated epidermis of wheat. Epidermal cells (EP) may elongate to 10 times their original length so that the mature cells are long and narrow. They are tubular, and the outer tangential wall arches out into the soil fabric, creating deep grooves (GR) between the cells. The cell surface is basically smooth with a few bacteria and persistent remnants from the cap-epidermal cell interface. One epidermal cell has produced a tubular root hair (RH). The grooves between epidermal cells often are sites of primary colonization by bacteria. (×3,700)

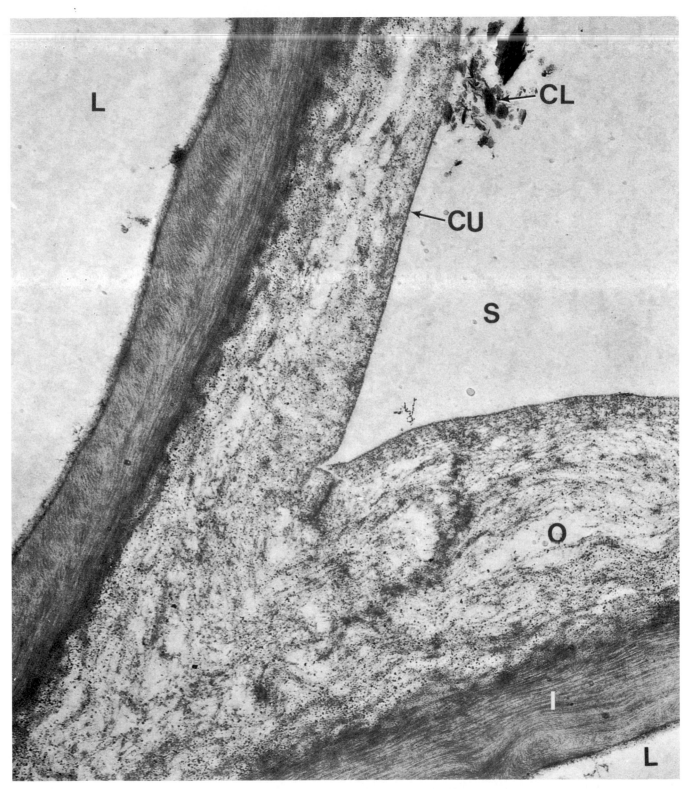

Fig. 30. Groove between two epidermal cells of wheat root. This section of the groove between two cells shows the three layers in the cell wall of the mature root epidermal cell. The outer surface is bound by a fine electron-dense cuticle (CU). This covers the granular primary wall (O) material, which in turn encloses the lamellate secondary wall (I). Clay minerals (CL) are attached to the cuticle. Unlike the primary cell wall material, which is continuous with and identical in structure and histochemistry to the middle lamella material, the cuticle does not extend into the radial wall. L marks the internal lumen of the epidermal cells; the surface external to the root makes contacts with soil (S) and clay. (×34,000) Reprinted, by permission, from R. C. Foster, The ultrastructure and histochemistry of the rhizosphere, New Phytol. 89:263-273, Plate 2, Fig. 1, ©1981.

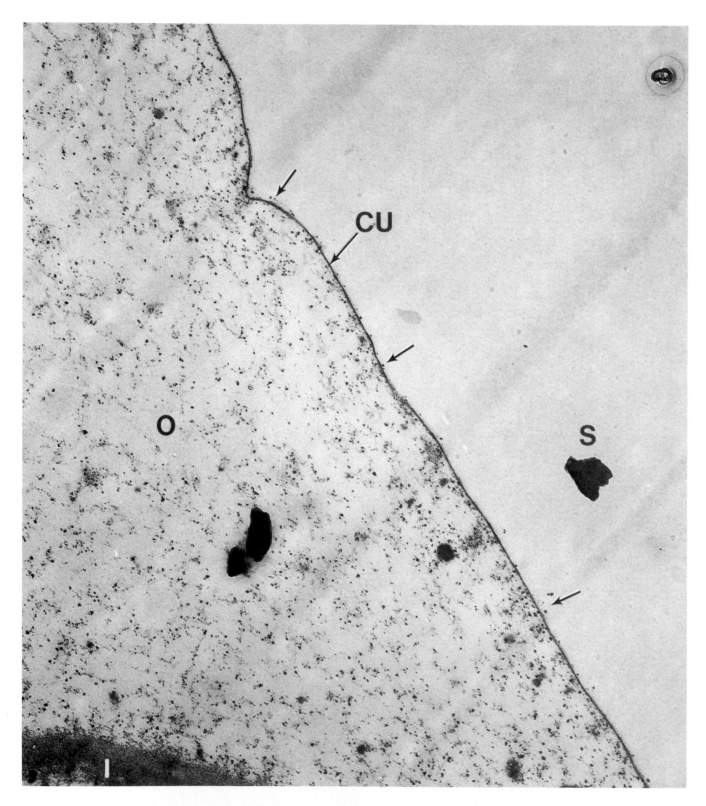

Fig. 31. Root surface details of cuticle and primary wall of pine. At higher magnification, primary wall material (O) appears to have an open, granular texture. This may partially be an artifact of preparation. Osmium tetroxide does not react with the carbohydrates that are the major component of the mucilage; only minor components, such as amino acids, lipids, and polyphenols, may be visible. Note the electron-dense cuticle (CU) with occasional patches of clay particles attached to its outer surface (arrows). I = inner layer of epidermal cell wall, S = soil. (×14,000)

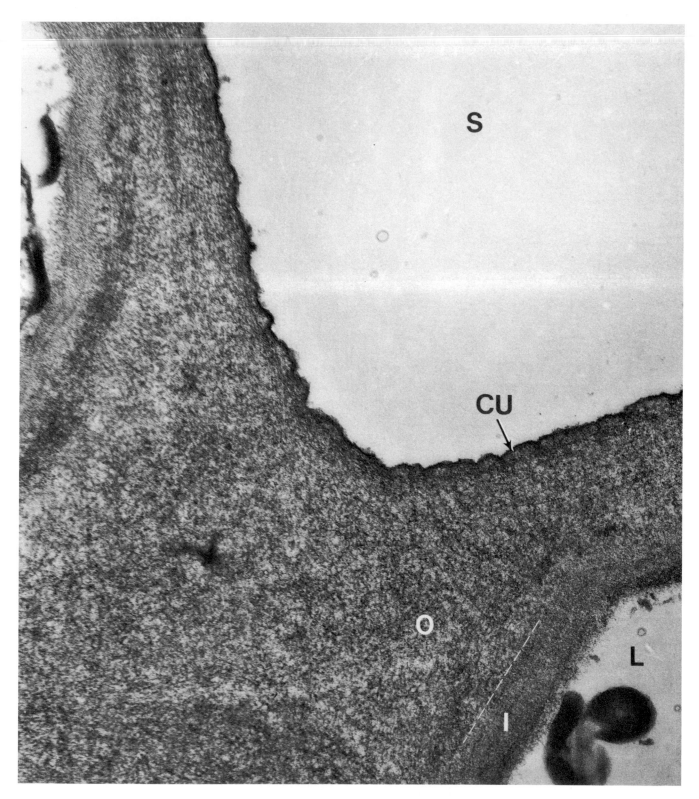

Fig. 32. Details of primary wall structure in pine. When the root is treated with lanthanum hydroxide, the acidic polysaccharides that are invisible after conventional fixation with osmium tetroxide are made electron-dense so that the primary wall appears to have a much denser texture and the boundary (– – –) between the primary (O) and secondary (I) walls is not so marked. The cuticle (CU) is also intensely stained, and as before, there are in places, traces of a trilamellate structure. L = lumen of the epidermal cell, S = soil. (Lanthanum stain, ×50,000) Reprinted, by permission, from R. C. Foster, The ultrastructure and histochemistry of the rhizosphere, New Phytol. 89:263-273, Plate 2, Fig. 3, ©1981.

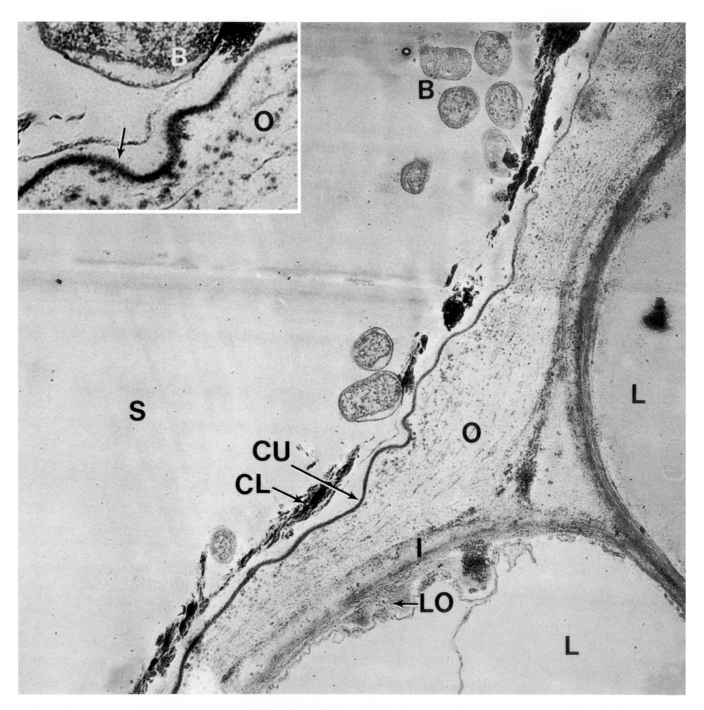

Fig. 33. Clover root. The cuticle (CU) is well developed in the family Fabaceae and may be the site of lectins involved in the host symbiont recognition process involving rhizobium bacteria (B). Clay (CL) particles are attached to the root surface, perhaps by remnants of cap mucilage, which are unstained. (×17,000) The inset shows that both the outer and inner surfaces of the cuticle are covered with small granules (arrows). I = inner layer of epidermal cell wall, L = lumen of epidermal cell, LO = lobe of inner wall layer, O = outer layer of epidermal cell wall, S = soil. (×48,000) Reprinted, by permission, from R. C. Foster, The fine structure of epidermal cell mucilages of roots, New Phytol. 91:727-740, Plate 2c, ©1982.

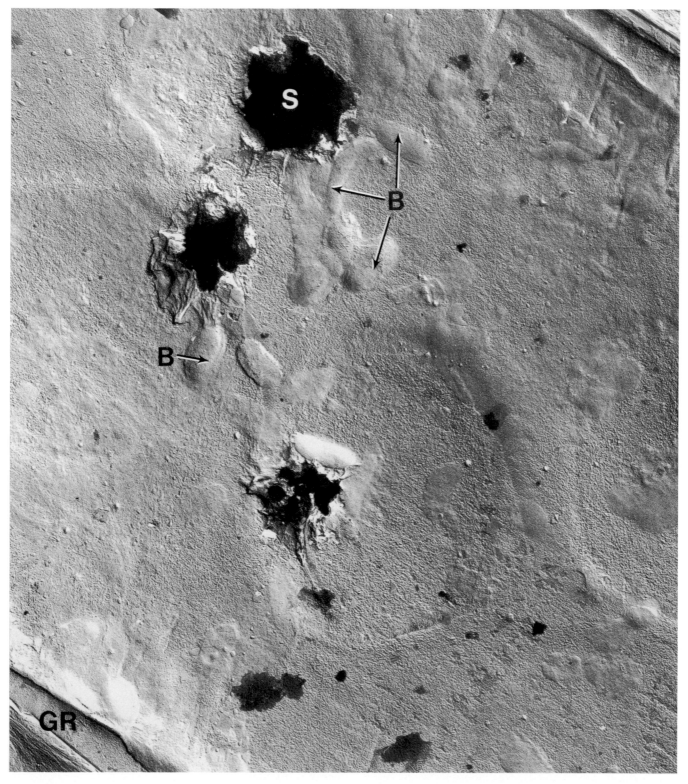

Fig. 34. TEM of a replica of the root surface of wheat. The root surface is covered with a fine granular material. Bacteria (B) are attached, with no sign of wall lysis at this stage. GR = groove between epidermal cells, S = soil. (×9,800)

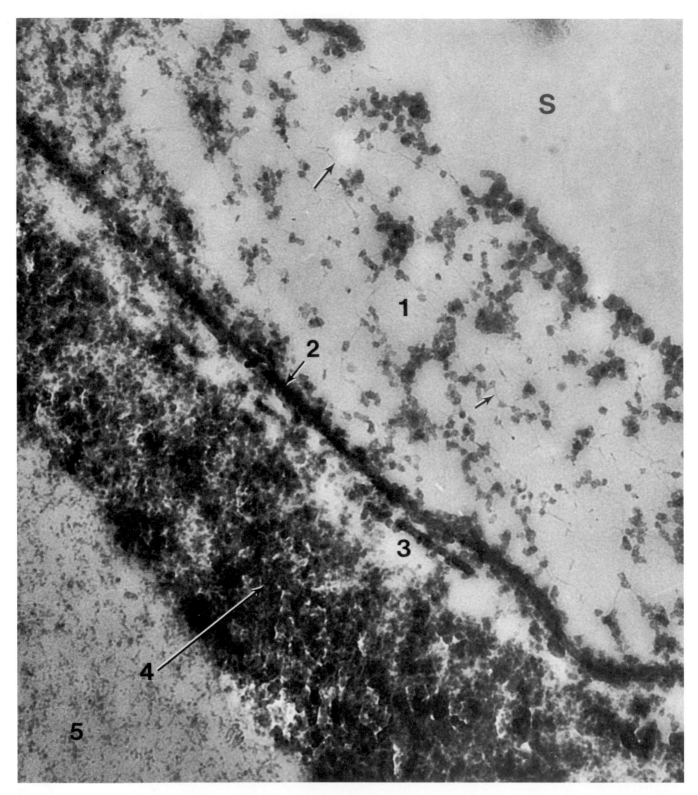

Fig. 35. Root-soil interface of wheat grown in the field. The epidermal mucilage on roots grown in the field under conditions of intermittent drying generally has a much more complex structure than that on plants grown in the more equable greenhouse conditions. In this TEM, the internal ultrastructure of the mucilage was investigated by binding the enzyme peroxidase to mucilage components with glutaraldehyde. The enzyme was then allowed to catalyze a reaction between diaminobenzidine and hydrogen peroxide to produce an insoluble polymer. The location of the polymer (and hence of the peroxidase and structure in the cell wall to which it was attached) was then demonstrated by staining it with osmium tetroxide. This reaction reveals that the mucilage is deposited in layers that differ in texture or composition. This TEM shows the outer five of seven layers in a mucilage of a wheat root sampled in summer at anthesis. S = soil, 1–5 = layers in mucilage. Arrows in layer 1 indicate microfibrils. (×64,000) Reprinted, by permission, from R. C. Foster, The fine structure of epidermal cell mucilages of roots, New Phytol. 91:727-740, Plate 3b, ©1982.

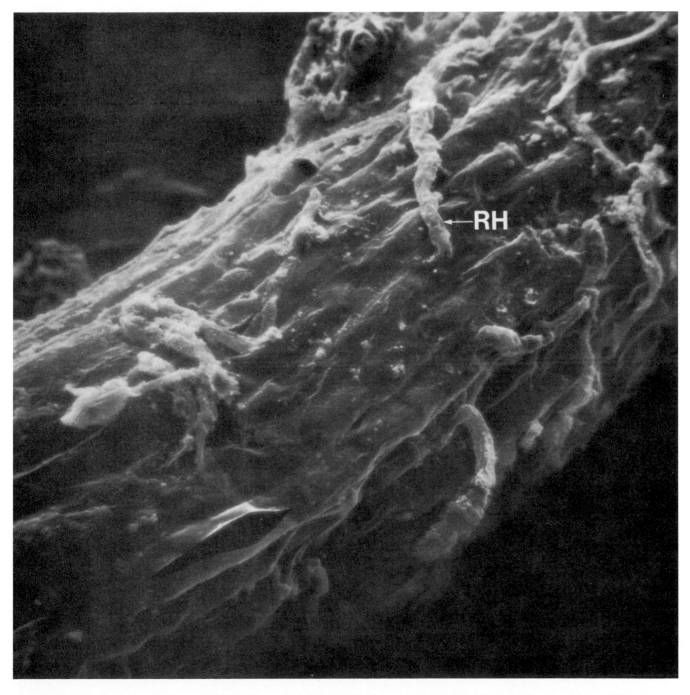

Fig. 36. Root hair zone of wheat. Behind the zone of elongation some epidermal cells produce root hairs (RH) that are 50–300 μm long. (×570)

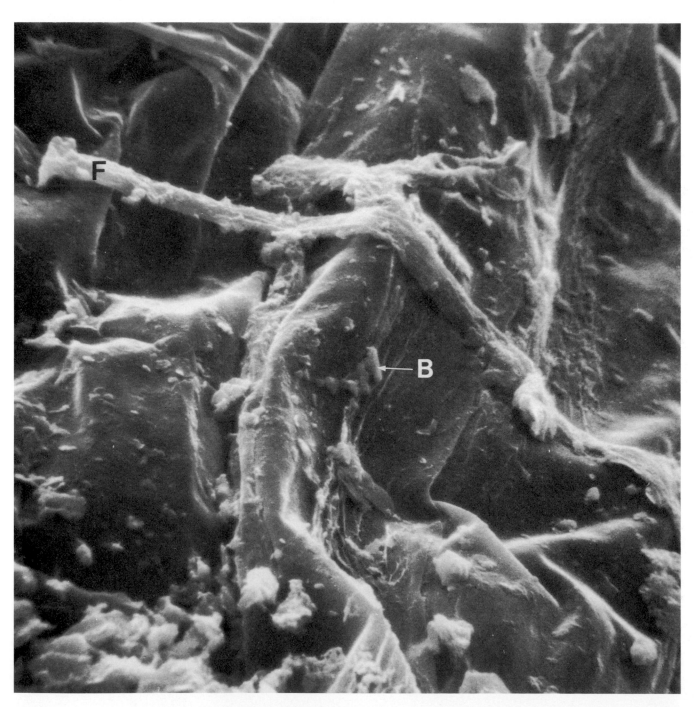

Fig. 37. Older root hair zone of wheat. Epidermal cells in the root hair zone are the first to become extensively colonized by bacteria (B) and fungi (F). (×2,700) Reprinted, by permission, from A. D. Rovira and R. Campbell, Scanning electron microscopy of microorganisms on the roots of wheat, Microb. Ecol. 1:15-23, Fig. 2A, ©1974, Springer-Verlag, New York.

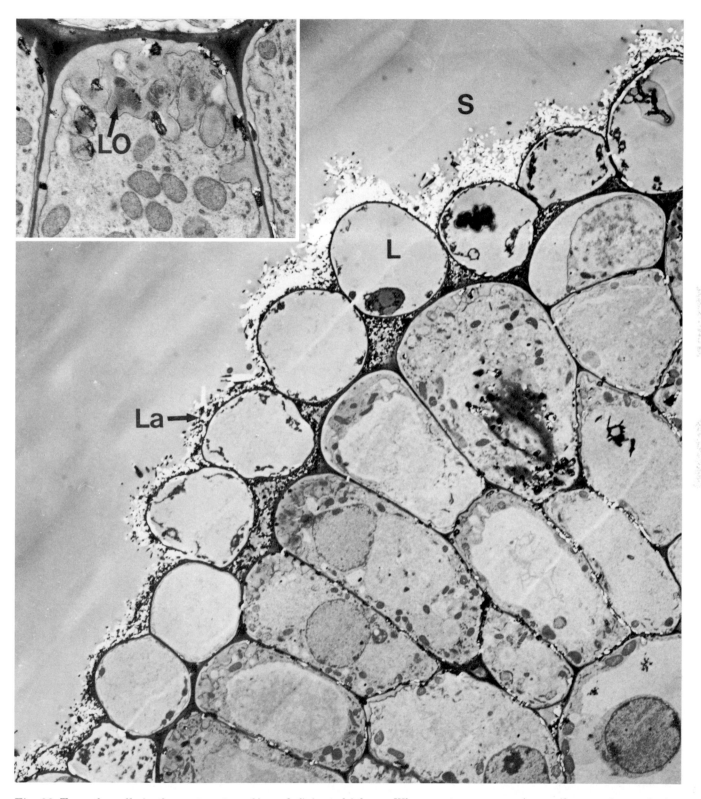

Fig. 38. Transfer cells in the root cortex of iron-deficient chickpea. When roots are grown in a soil or nutrient solution deficient in iron, many of the epidermal and subepidermal cells are differentiated as transfer cells. (×4,500) The inset shows the curious fingerlike lobes (LO) that grow into the peripheral cytoplasm. L = lumen of epidermal cell, La = lanthanum precipitate to show mucilage, S = soil. (×11,000)

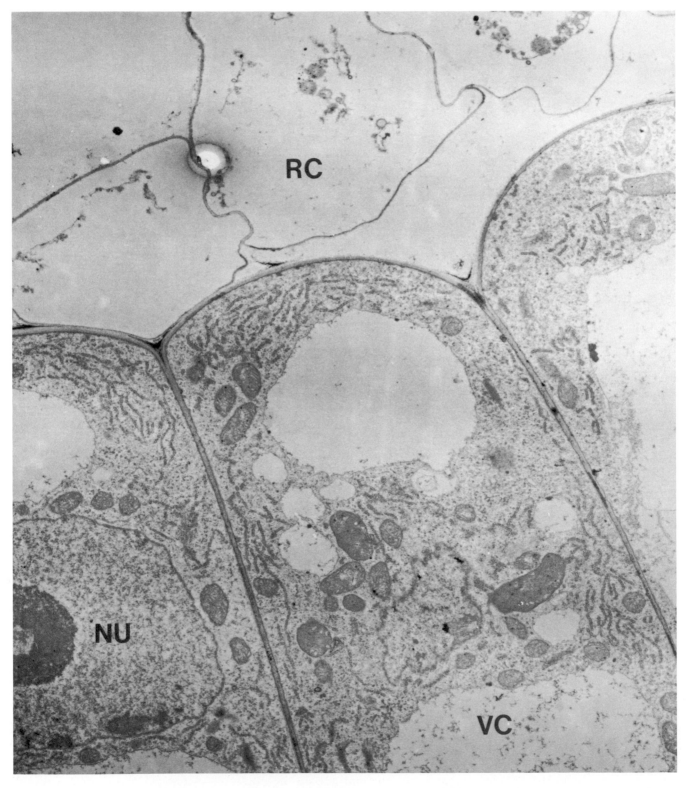

Fig. 39. Epidermal cells of chickpea roots with adequate iron nutrition. Roots provided with adequate iron do not produce transfer cells in the root cortex. NU = nucleus, RC = root cap cell, VC = vacuole. (×10,200)

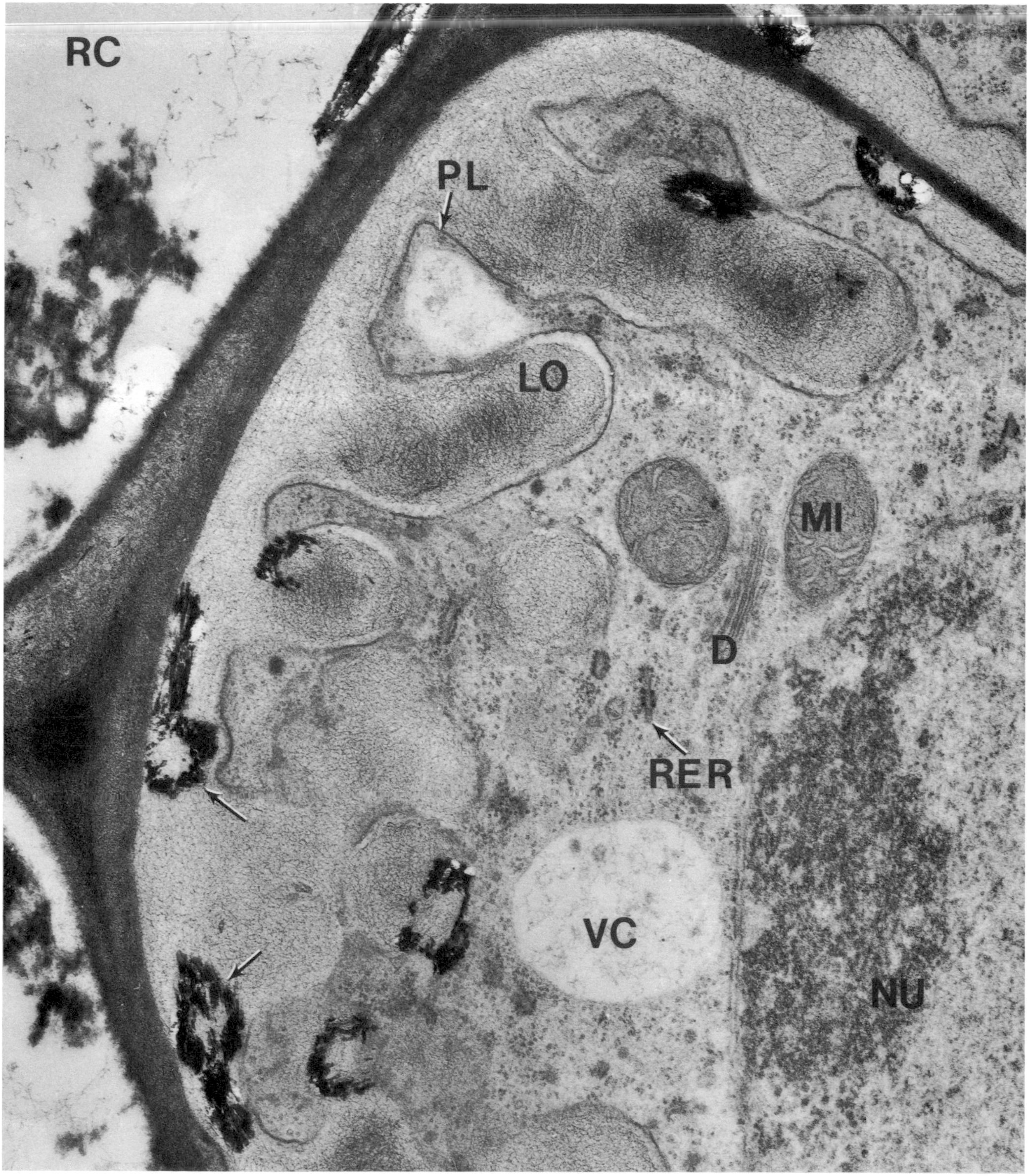

Fig. 40. Details of the structure of cortical transfer cells in chickpea. In transfer cells, the fingerlike processes are produced by the inner layer of the outer tangential cell wall. Each process is enclosed by the cell membrane or plasmalemma (PL), and the processes greatly increase the area of contact between the wall free space and the plasmalemma. The surrounding cytoplasm is filled with organelles, mitochondria (MI), dictyosomes (D), and rough endoplasmic reticulum (RER). Note the deposits of lanthanum phosphate that were used to mark the limits of the root free space (arrows). This apparently extends into the processes. LO = lobe of inner cell wall, NU = nucleus, RC = root cap cell, VC = vacuole. (×47,000)

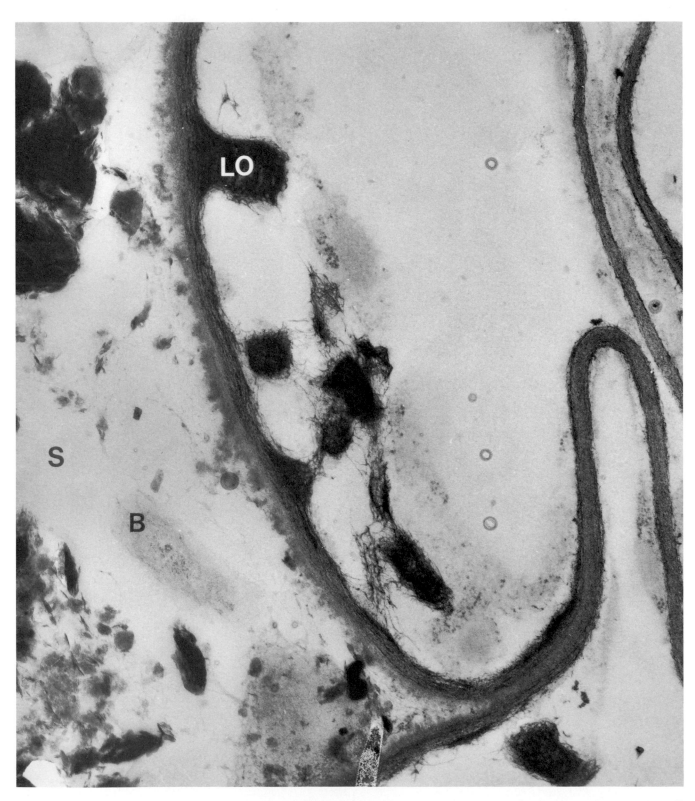

Fig. 41. Old epidermal transfer cells in chickpea. When ruthenium red is included in the fixative, the transfer cell processes stain intensely, indicating the presence of substituted polysaccharides; these may be implicated in the ion exchange properties of the lobes. In older cells the lobes become highly branched. B = bacterium, LO = lobe of inner cell wall, S = soil. (×62,000)

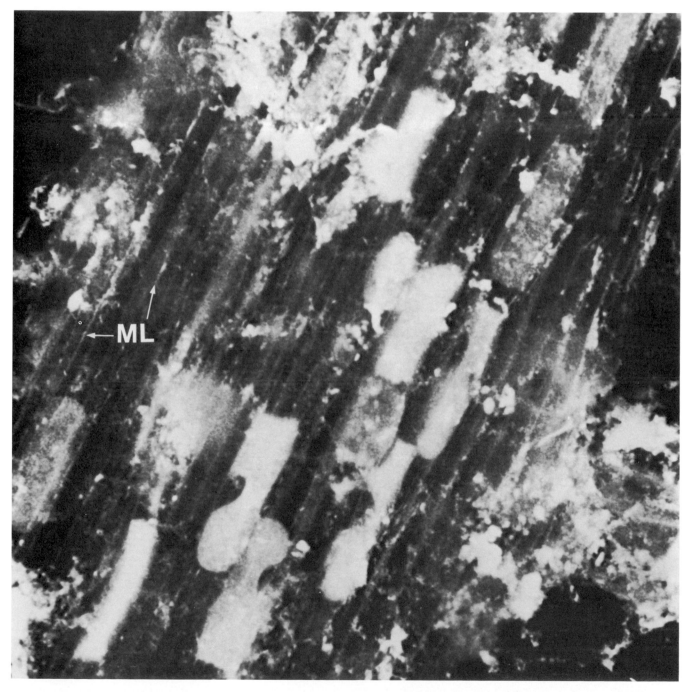

Fig. 42. Young root of wheat. Behind the root hair zone, the epidermal cells autolyse. In this preparation, the root was fixed with osmium tetroxide and examined in backscatter SEM. Only living cells that contained osmium-stained cytoplasm returned a signal (white); lysed epidermal cells appear dark and greatly outnumber the live cells. No pattern could be detected in the order in which cells lysed. ML = osmium-reactive middle lamella of radial wall. (×300)

Fig. 43. Young wheat root. Quartz (sand) grains (Q) and clay particles (CL) become embedded in the root surface as the root penetrates the soil (×3,300) Reprinted, by permission, from A. D. Rovira and R. Campbell, Scanning electron microscopy of microorganisms on the roots of wheat, Microb. Ecol. 1:15-23, Fig. 3B, ©1974, Springer-Verlag, New York.

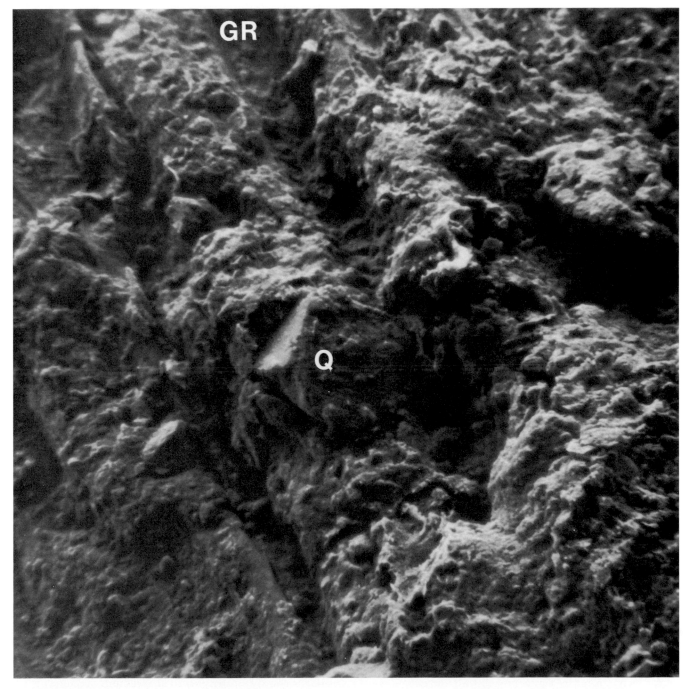

Fig. 44. Wheat root. Older parts of the root are coated with soil minerals embedded in the mucilage of the primary wall. At this stage the mucilage has not penetrated far into the soil fabric so that the outline of the cells and the deep grooves (GR) between them are still visible. Q = quartz. (×1,200)

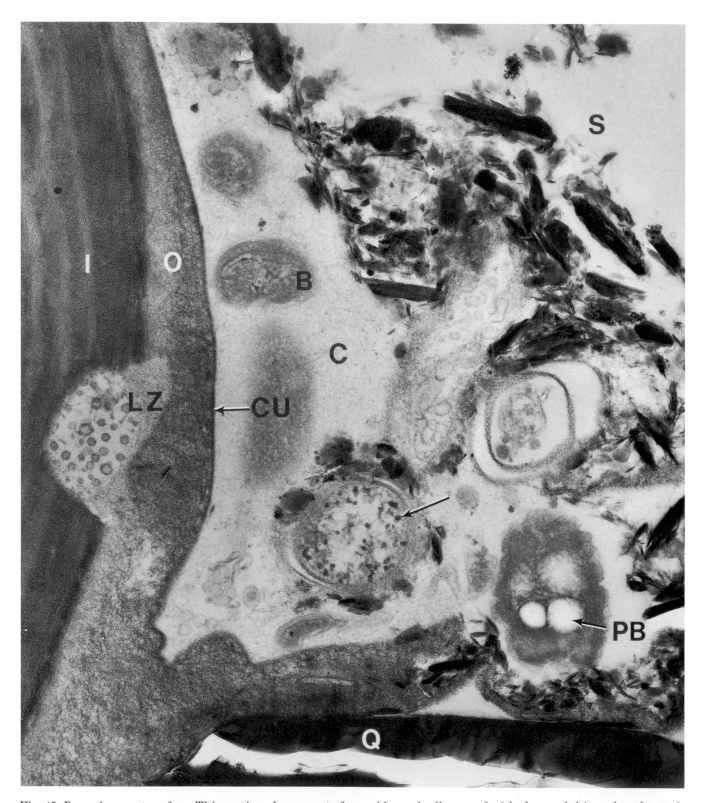

Fig. 45. *Paspalum* root surface. This section shows part of an epidermal cell covered with clay and rhizosphere bacteria (B). A quartz particle (Q) has ruptured the cuticle (CU) and has become embedded in the primary wall mucilage (O), which has been made visible by the periodic acid-thiosemicarbazide silver proteinate (PATSP) stain for neutral polysaccharides. This stain also shows the herringbone arrangement of the microfibrils in the secondary wall (I) of the epidermal cell. Glycogen (arrow) in rhizosphere bacteria was also stained by the PATSP, but the polyhydroxybutyrate (PB) was not, and the bacterial cell wall and capsule materials (C) were only slightly stained. Previous sections of this epidermal cell (not shown) indicated that the small profiles in a lysis zone (LZ) in the cell wall were attached to a bacterium. (×55,000) Reprinted, by permission, from R. C. Foster, The ultrastructure and histochemistry of the rhizosphere, New Phytol. 89:263-273, Plate 3, Fig. 1, ©1981.

60

Fig. 46. Older wheat root. With the rupture of the cuticle, mucilage escapes from the root surface into the soil and coats the nearby soil components. Quartz grains (Q) are now covered with clay (CL), stuck onto their surfaces by the mucilage. Mucilage also holds bacteria to the surfaces of the soil minerals. (×3,000)

Fig. 47. Root surface of *Paspalum*. Near the cell lumen (L), the mucilage (M) is very dense but toward the root surface it is evidently more fluid so that, when soil minerals press into it, the mucilage flows around and embeds the soil particles penetrating into the soil fabric beyond (arrows). I = inner layer of epidermal cell wall, O = outer layer of epidermal cell wall, S = soil. (Lanthanum stain, ×14,000)

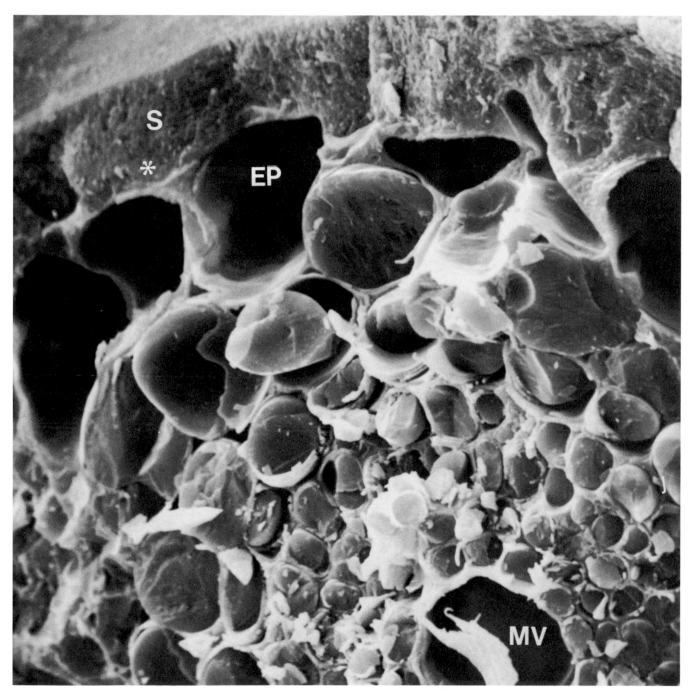

Fig. 48. Root of wheat. The extent to which the mucilage penetrates the soil (S) is demonstrated by this SEM of a root and rhizosphere. The root and soil have been frozen in liquid nitrogen slush, snapped transversely, and shadowed with heavy metal. The soil held at the root surface is almost equal in thickness to the diameter of the nearby cortical cells. This section is not near the root cap because the metaxylem vessel (MV) is fully differentiated. The region marked * is shown at higher magnification in Fig. 49. EP = epidermal cell. (×2,000) Reprinted, by permission, from R. Campbell and R. Porter, Low temperature scanning electron microscopy of microorganisms in soil, Soil Biol. Biochem. 14:241-245, Fig. 7, ©1982, Pergamon Press, Oxford.

Fig. 51. *Paspalum* rhizosphere. A quartz particle (Q) lies at the surface of the more highly polymerized inner mucilage. The more fluid outer mucilage has embedded the nearby soil fabric to a distance of several microns. A small colony of bacteria (B) has developed in the soil fabric, and isolated bacteria are scattered through the soil. A fungal hypha (F) has grown in the mucilage in the groove between two epidermal cells. I = inner layer of epidermal cell wall, L = epidermal cell lumen, O = outer layer of epidermal cell wall. (Periodic acid-thiosemicarbazide silver proteinate stain, ×15,000) Reprinted, by permission, from R. C. Foster, The ultrastructure and histochemistry of the rhizosphere, New Phytol. 89:263-273, Plate 4, Fig. 1, ©1981.

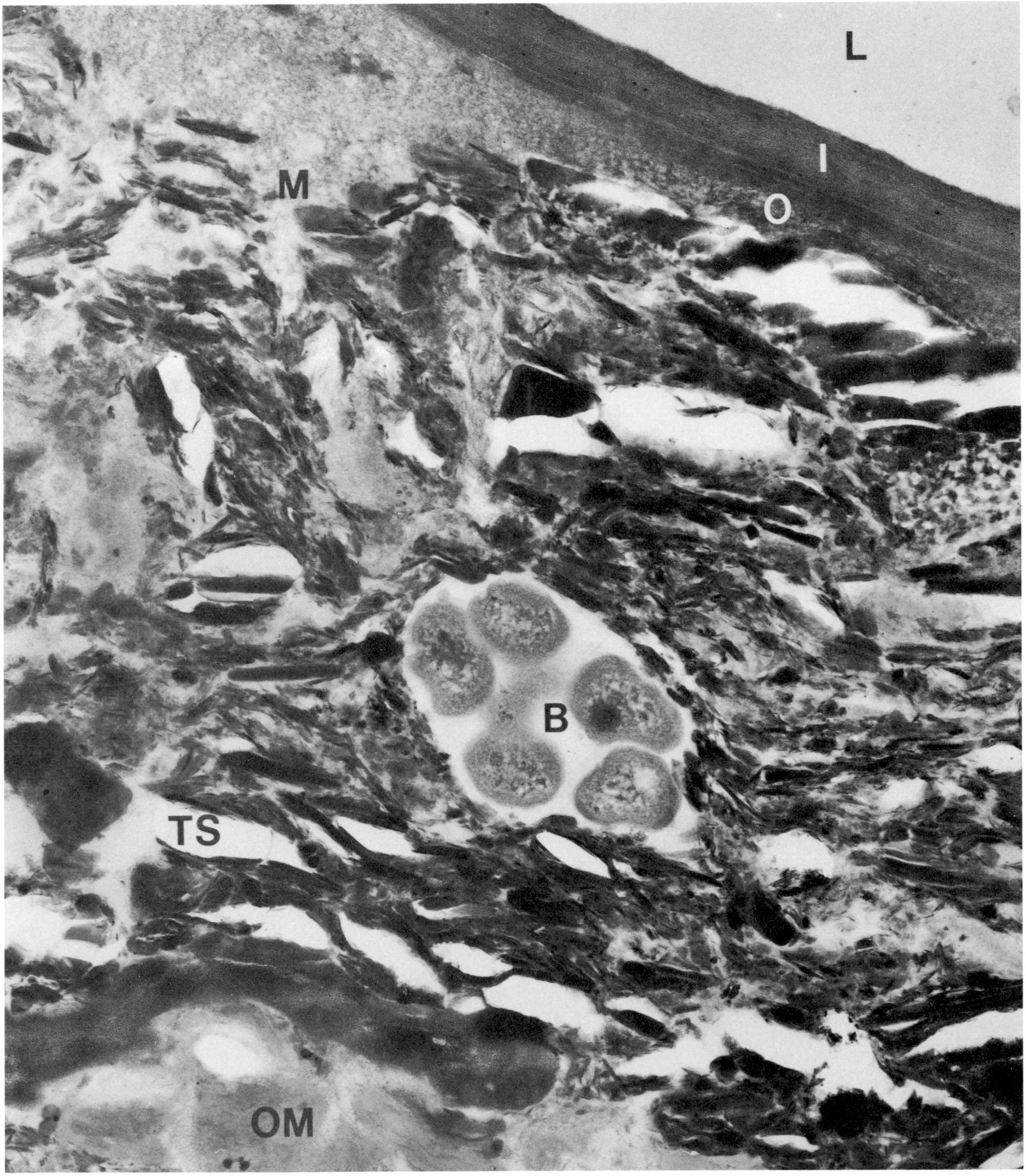

Fig. 52. *Paspalum* rhizosphere. At higher magnifications, the mucilage can be seen to have embedded the whole soil fabric so that voids smaller than 1 μm across are filled with mucilage (M). The capsule of the bacteria (B) has not stained. There has been some movement of clay particles during ultramicrotomy causing small electron-transparent tears (TS) in the section. I = inner layer of epidermal cell wall, L = epidermal cell lumen, O = outer layer of epidermal cell wall, OM = organic matter. (Periodic acid-thiosemicarbazide, ×33,000)

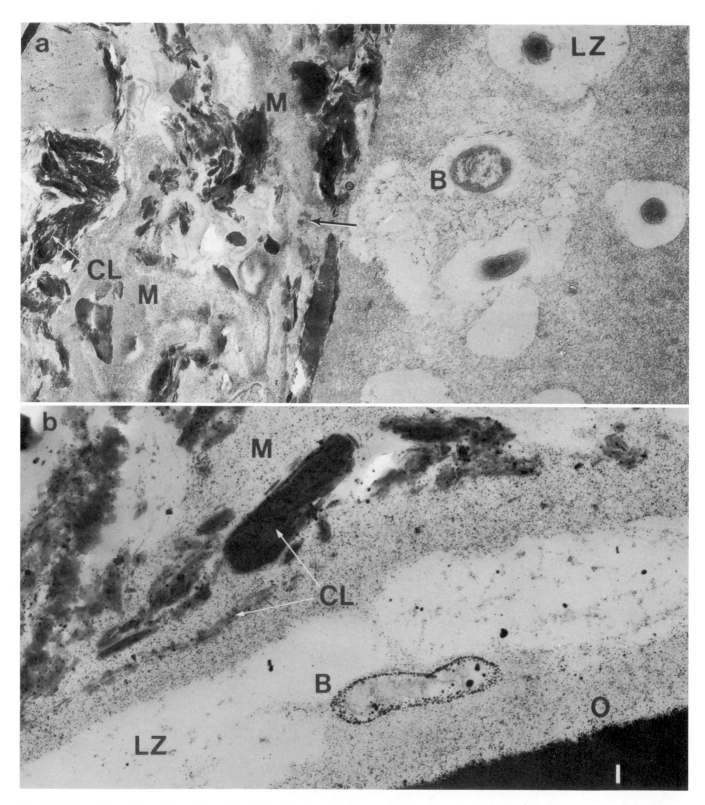

Fig. 53. *Paspalum* rhizosphere. Two histochemical methods were used to demonstrate the fluid outer mucilage (M). By both methods, the micrographs show that the mucilage has been invaded by bacteria (B), which lie in the lysis troughs due to enzymatic action. The mucilage flows around and embeds the clay fragments (CL) (arrows). In **a,** lanthanum hydroxide was used to stain the substituted carbohydrates. (×13,000) Reprinted, by permission, from R. C. Foster, Ultramicromorphology of some South Australian soils, Plate 3, in Modification of Soil Structure, W. W. Emerson, R. D. Bond, and A. R. Dexter, eds., ©1978, John Wiley & Sons, New York. In **b,** the periodic acid-silver methanamine has stained neutral carbohydrates. (×55,000) Reprinted, by permission, from R. C. Foster and J. K. Martin, In situ analysis of soil components of biological origin, Plate 8a, in Soil Biochemistry, vol. 5, E. A. Paul and J. N. Ladd, eds., ©1981, Marcel Dekker, Inc., New York. I = inner layer of epidermal cell wall, LZ = lysis zone, O = outer layer of epidermal cell wall.

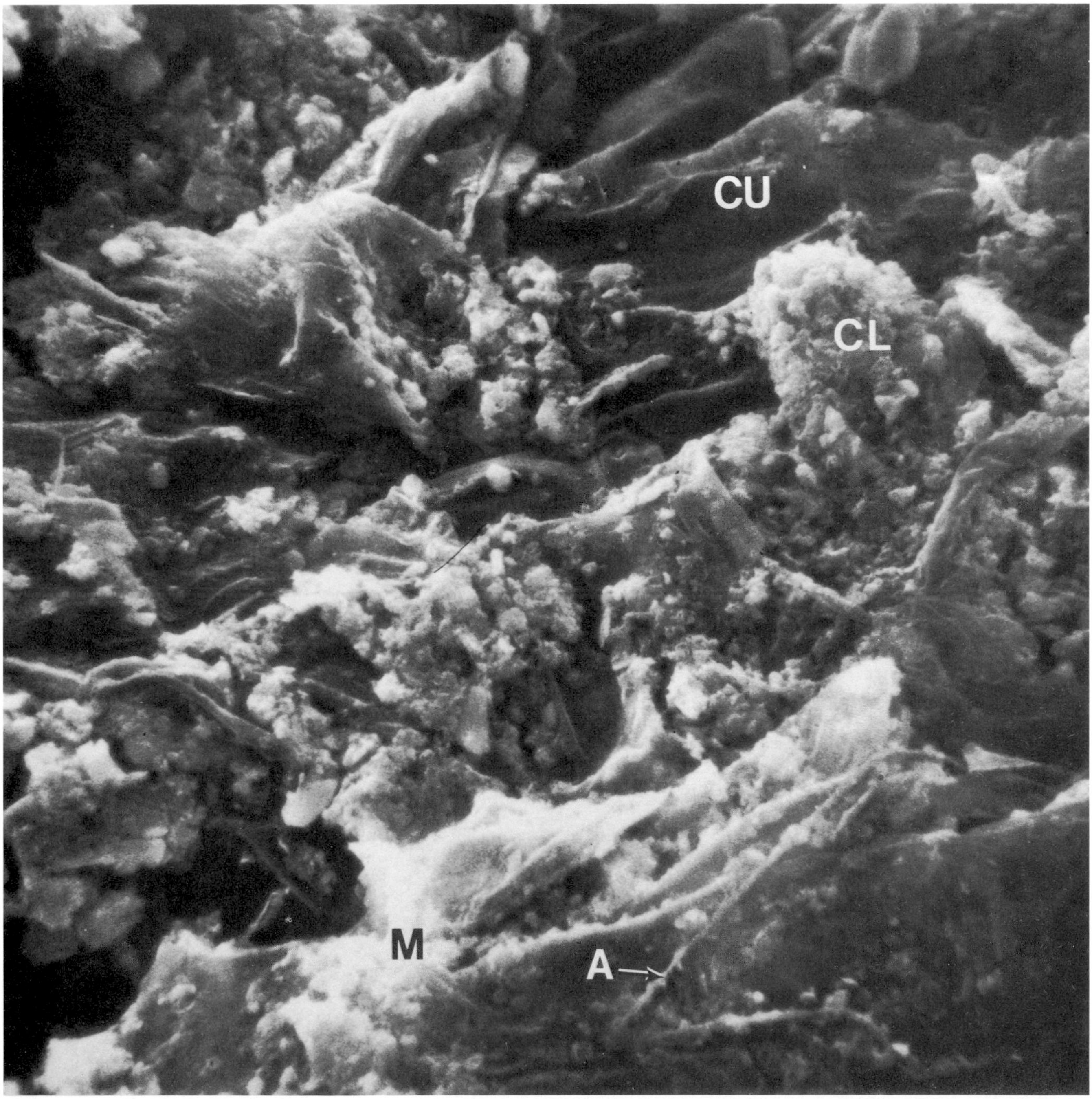

Fig. 54. Wheat root surface. The complexity of the root surface greatly increases with the escape of the nutrient-rich mucilage (M). Except where the cuticle (CU) remains unruptured, the cell surface is covered with minerals (CL), bacteria, and actinomycetes (A). (×1,200)

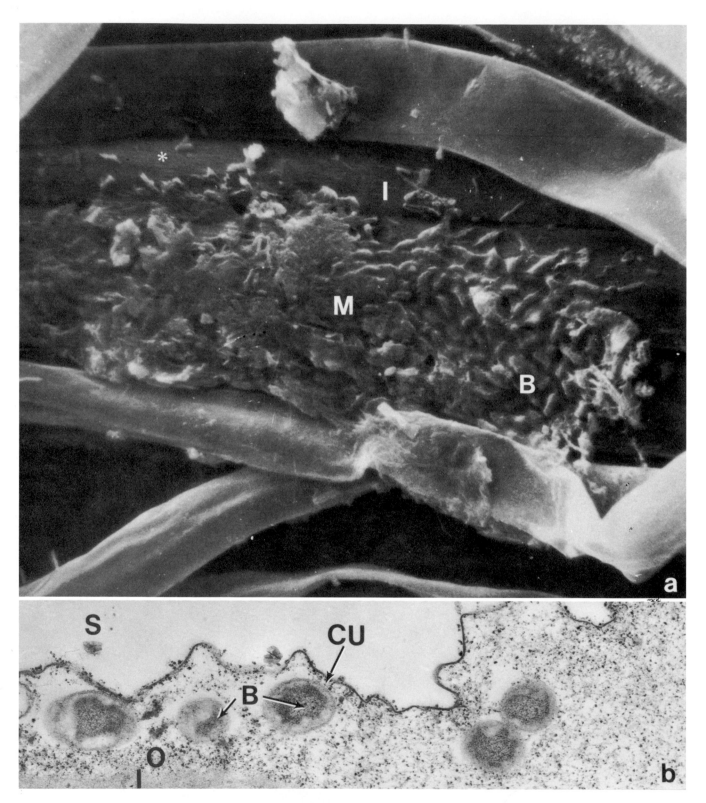

Fig. 55. Bacteria in wheat roots. **a,** SEM. After the cuticle (CU) has ruptured, soil bacteria invade and lyse the mucilage; they can then be distinguished under the cuticle covering the mucilage. When the mucilage is ruptured, the bacteria can be seen on the root surface. At M, the bacteria are obscured by mucilage, but at B, they can be seen through the cuticle. At * the cuticle has broken down, exposing the individual microbial cells. (×2,850) Reprinted, by permission, from A. D. Rovira and R. Campbell, Scanning electron microscopy of microorganisms on the roots of wheat, Microb. Ecol. 1:15-23, Fig. 4a, ©1974, Springer-Verlag, New York. **b,** TEM of bacteria below the cuticle. Bacteria (B) may divide in the mucilaginous outer layer of the cell wall (O), enclosed by the cuticle. The absence of a lysis zone suggests that the bacteria are utilizing soluble exudates from the cells below. I = inner layer of epidermal cell wall, S = soil. (×37,000)

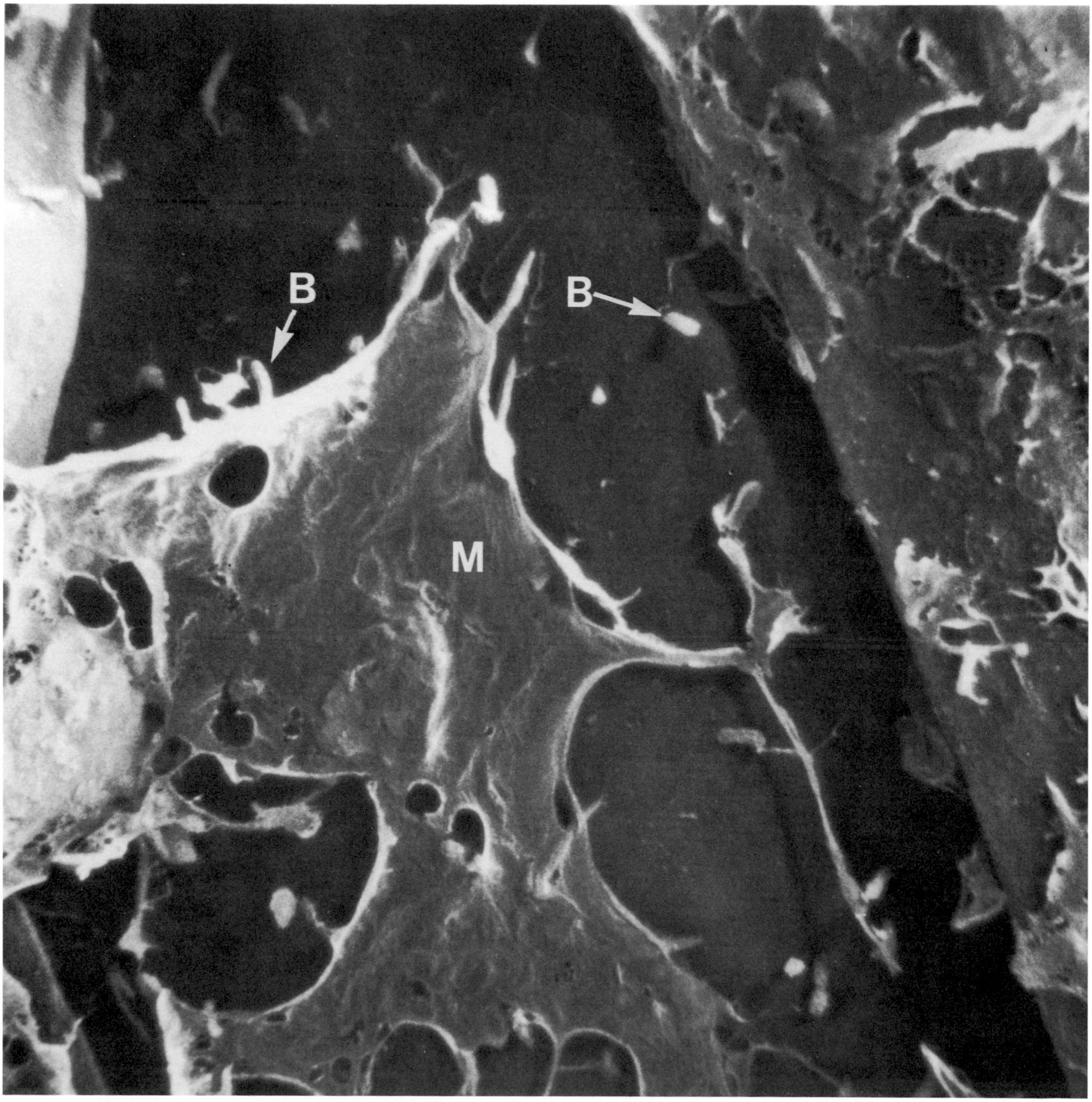

Fig. 56. Old wheat root. Following extensive microbial lysis of the mucilage (M), the cuticle and remains of the mucilage fold back, revealing the clear outlines of bacteria (B) attached to the secondary wall of the epidermal cells. (×5,900) Reprinted, by permission, from A. D. Rovira and R. Campbell, A scanning electron microscope study of the interactions between microorganisms and *Gaeumannomyces graminis* (syn. *Ophiobolus graminis*) on wheat roots, Microb. Ecol. 2:177-185, Fig. 4, ©1975, Springer-Verlag, Inc., New York.

Fig. 57. Old wheat root. With the removal of the mucilage, microfibrils of the primary cell wall contact the surfaces of the minerals (arrows). These fibrils may aid in the transfer of ions from the surface of the clay into the root free space. L = lumen. (×108,000) Reprinted, by permission, from R. C. Foster, The fine structure of epidermal cell mucilages of roots, New Phytol. 91:727-740, Fig. 4, ©1982.

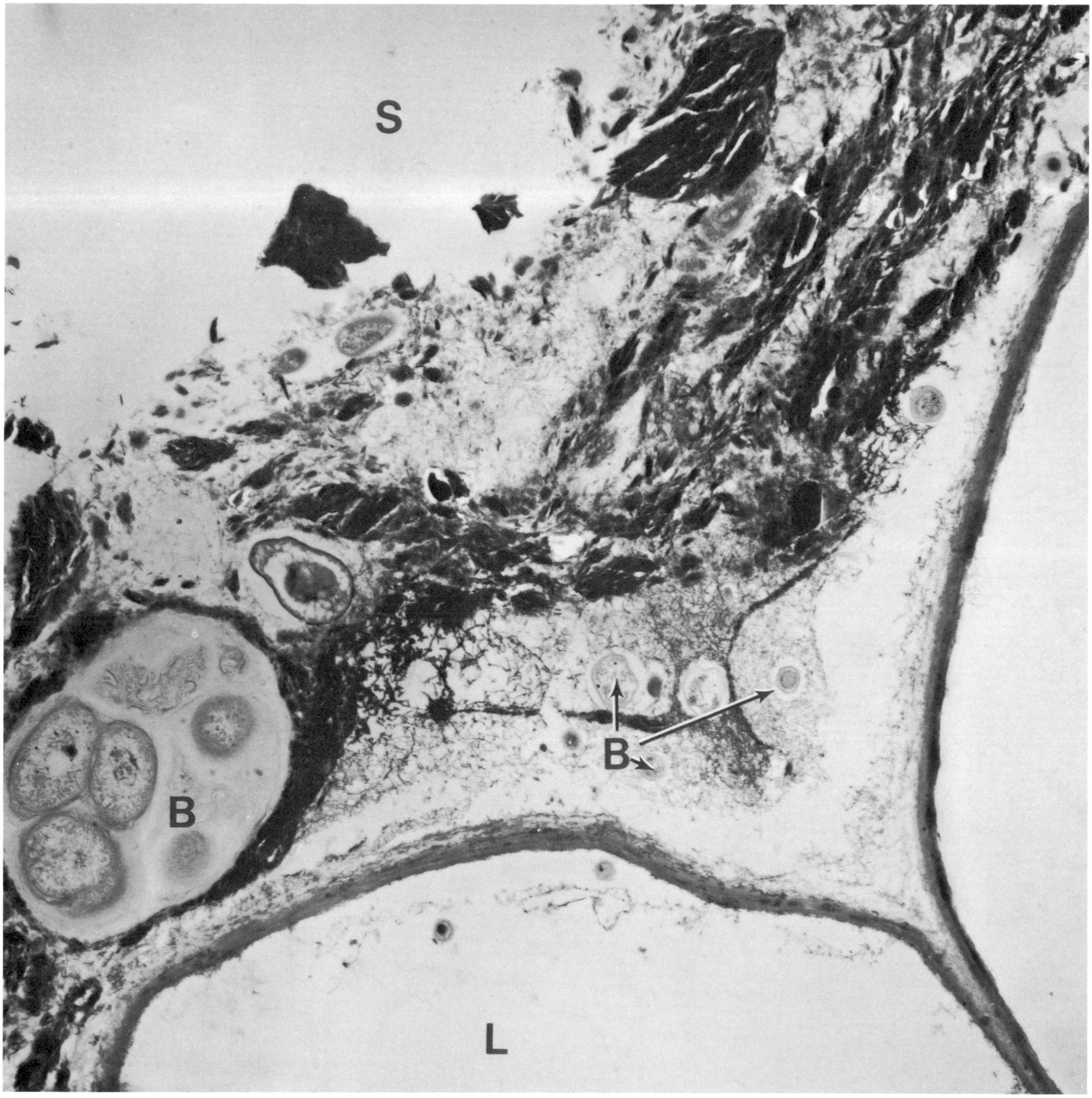

Fig. 58. Old rice root. Similar lysis of mucilage occurs in rhizosphere of flooded rice. Colonies of bacteria (B) can be seen both in the rhizosphere and embedded in the mucilage at the root surface. L = lumen of epidermal cell, S = soil. (×15,000) Reprinted, by permission, from R. C. Foster, The ultrastructure and histochemistry of the rhizosphere, New Phytol. 89:263-273, Plate 5, Fig. 2, ©1981.

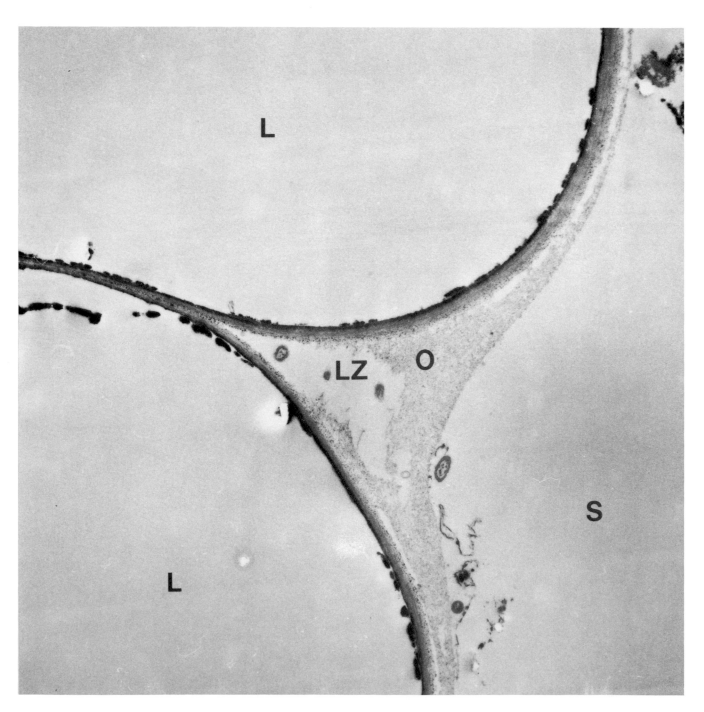

Fig. 59. Rhizoplane of pine. When the root is washed, only the highly polymerized inner mucilage remains at the root surface. This is less than 1 μm thick over the bulk of the cell wall but may almost fill the grooves between the cells. Most exudates probably escape from the intercellular spaces; consequently most bacteria on the root surface are found mainly in the grooves between the cells. L = epidermal cell lumen, LZ = lysis zone, O = outer layer of epidermal cell wall, S = soil. (×14,000)

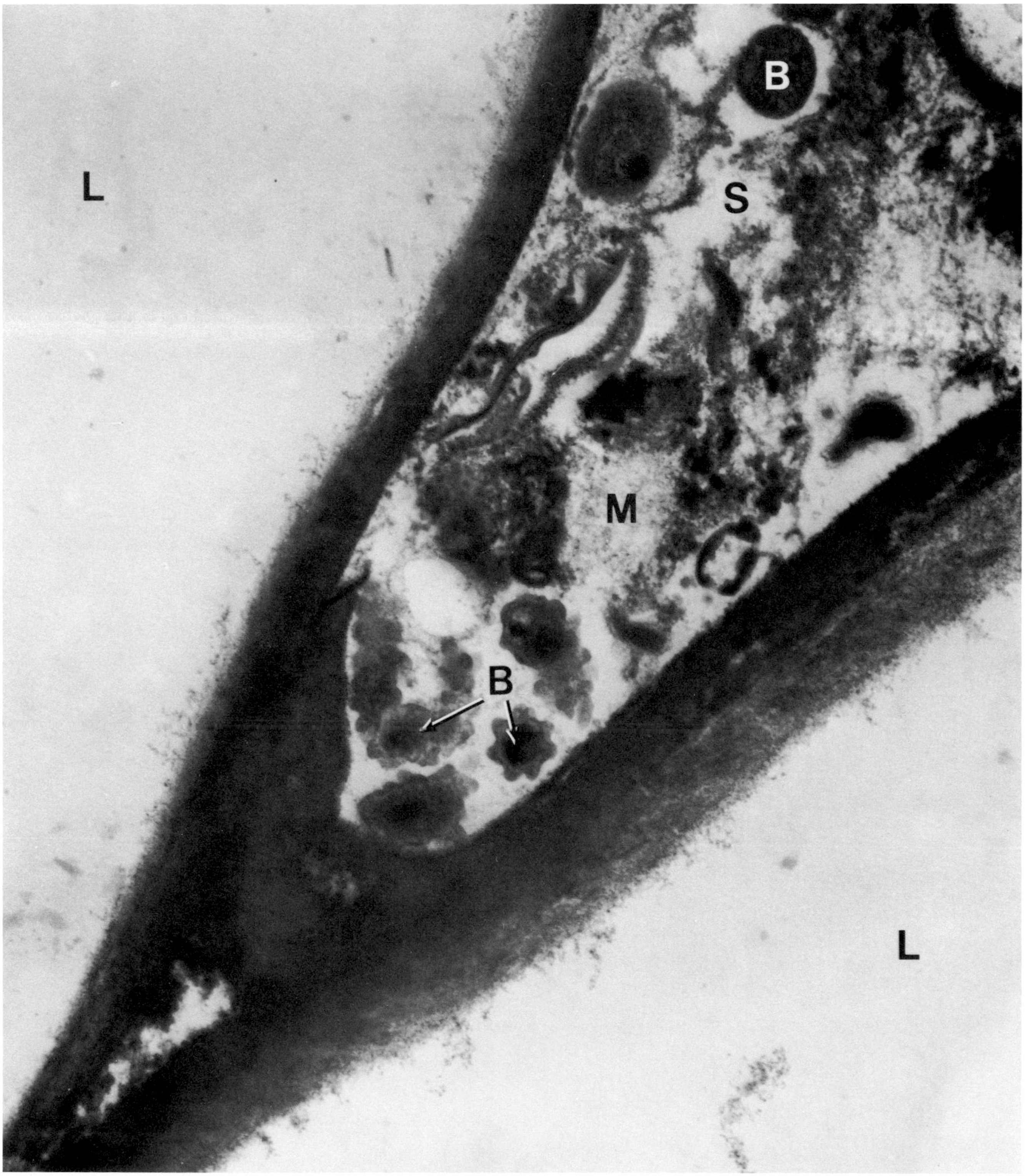

Fig. 60. Bacteria in the grooves between epidermal cells of pine. Large populations of bacteria (B) build up in the mucilage-filled grooves between the epidermal cells. In many species, the grooves contain unusual lobed bacteria that are uncommon elsewhere in the rhizosphere and that may be part of a specific rhizoplane microflora. L = epidermal cell lumen, M = mucigel, S = soil. (×27,000)

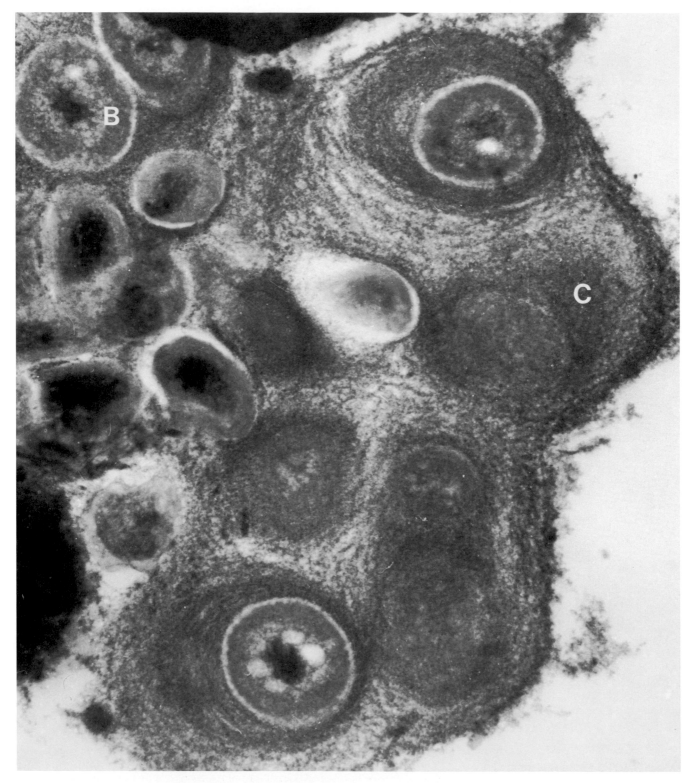

Fig. 61. Bacteria in a pine rhizosphere. The substituted carbohydrates in the capsule materials (C) of rhizosphere bacteria (B) often stain intensely with lanthanum hydroxide. Sometimes the capsule fibrils are deposited in concentric layers. (×40,000)

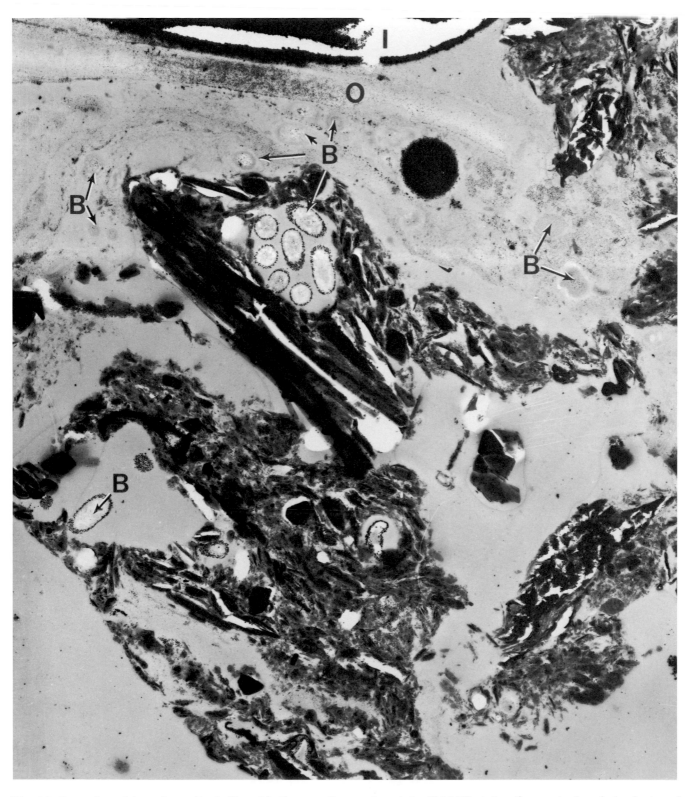

Fig. 62. *Paspalum* rhizosphere. Periodic acid-silver methanamine stain (PAMS) stains the neutral carbohydrates of the rhizosphere, including the cell walls of the bacteria. Capsules surrounding bacterial cells do not stain with PAMS. Colonies of bacteria build up in the rhizosphere, and many are enclosed in the clay fabric. Note the large number of bacteria (B) at the rhizoplane. I = inner and O = outer layers of the epidermal cell wall. (×5,000) Reprinted, by permission, from R. C. Foster, The ultrastructure and histochemistry of the rhizosphere, New Phytol. 89:263-273, Plate 5, Fig. 1, ©1981.

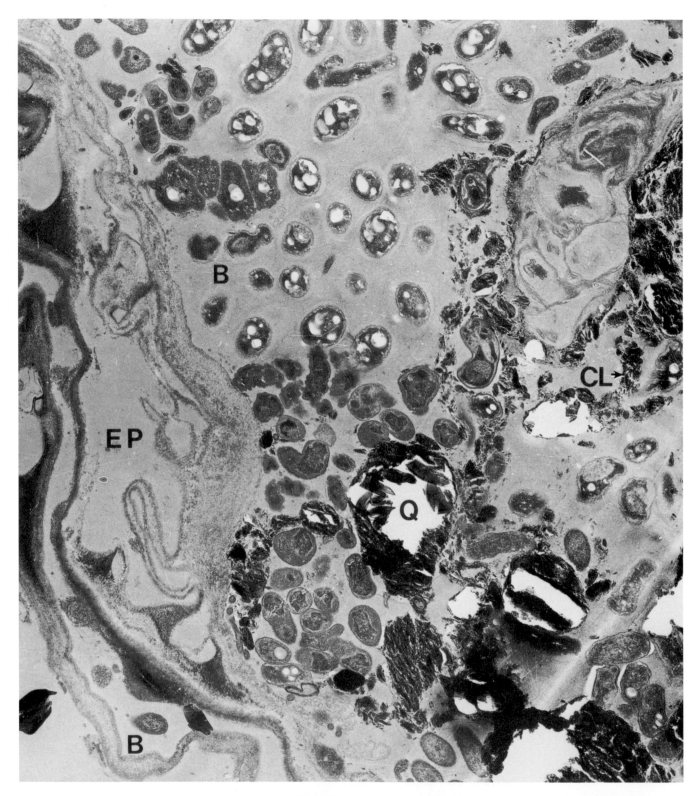

Fig. 63. Mature rhizosphere from clover. By anthesis, the outer cortex of the roots has become crushed and the epidermal cells (EP) become distorted with convoluted and partly lysed outer cell walls. With cell lysis, large populations of bacteria (B) now occur in the inner rhizosphere. Many morphological forms can be distinguished, and the various types often occur in small discrete colonies, many of which have a distinctive type of capsule material. CL = clay, Q = quartz grain. (×10,000) Reprinted, by permission, from R. C. Foster and A. D. Rovira, The ultrastructure of the rhizosphere of *Trifolium subterraneum* L., Fig. 1, in Microbial Ecology, M. W. Loutit and J. A. R. Miles, eds., ©1978, Springer-Verlag, Berlin, Heidelberg.

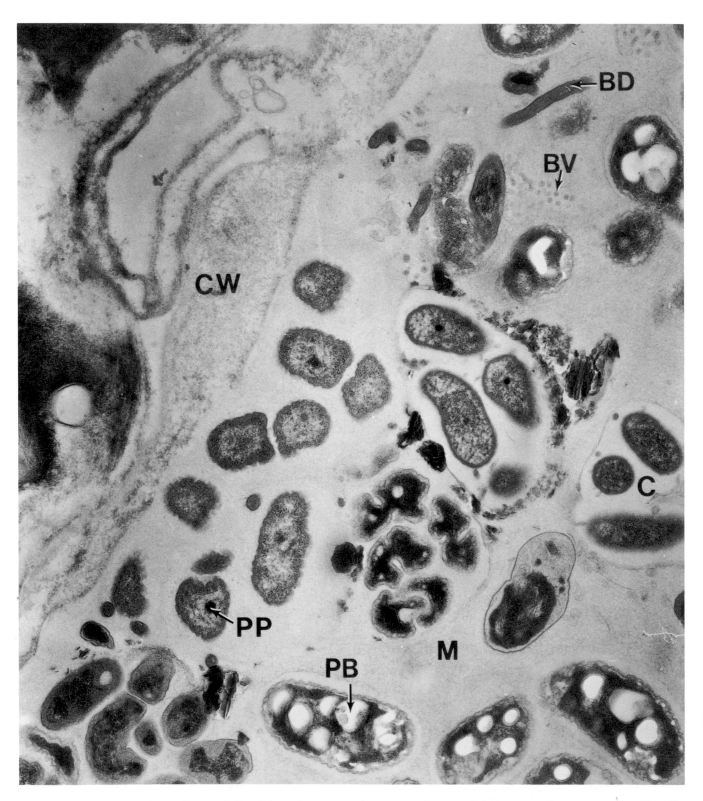

Fig. 65. Clover rhizosphere. Some of the rhizosphere organisms are of unusual form; some have smooth cell walls, others have lobed walls. Here the rhizosphere has been invaded by microbial parasites. The elongated profiles are the predatory bacterium, *Bdellovibrio* (BD), which is associated with several types of bacteria. Also present are bacteriophages (BV) recognized by their hexagonal heads with electron-dense cores. C = capsule material, CW = epidermal cell, M = mucigel, PB = polyhydroxybutyrate, PP = polyphosphate granules. (×25,000) Reprinted, by permission, from R. C. Foster and A. D. Rovira, The ultrastructure of the rhizosphere of *Trifolium subterraneum* L., Fig. 1, in Microbial Ecology, M. W. Loutit and J. A. R. Miles, eds., ©1978, Springer-Verlag, Berlin, Heidelberg.

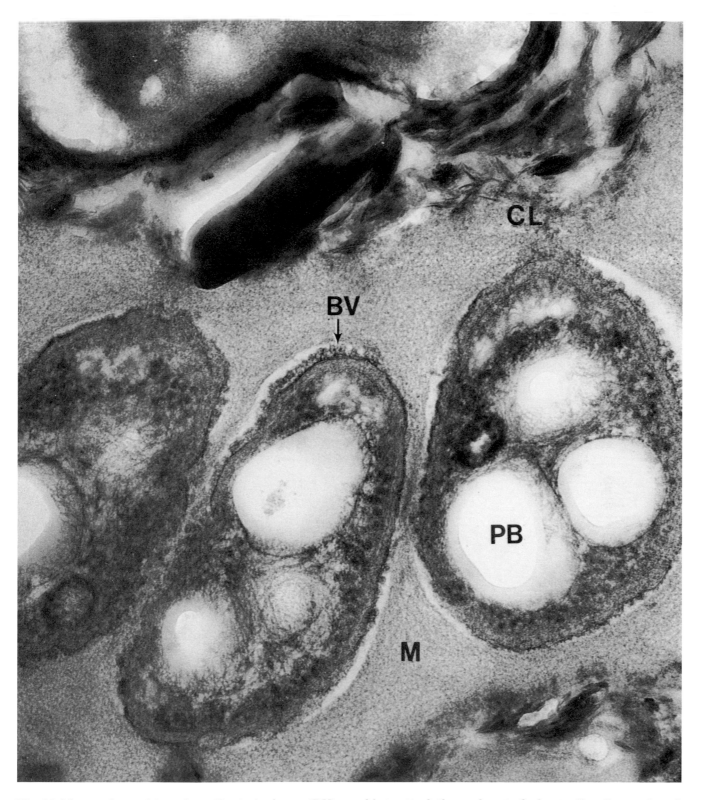

Fig. 66. Mature clover rhizosphere. Bacteriophages (BV) are able to attach themselves to the host cell wall even though the bacteria are enclosed in mucigel (M). Many virus particles have injected their DNA into the host. But some still contain an electron-dense core. Note the large deposits of polyhydroxybutyrate (PB) in the host cells. CL = clay. (×31,000) Reprinted, by permission, from R. C. Foster and A. D. Rovira, The ultrastructure of the rhizosphere of *Trifolium subterraneum* L., Fig. 23, in Microbial Ecology, M. W. Loutit and J. A. R. Miles, eds., ©1978, Springer-Verlag, Berlin, Heidelberg.

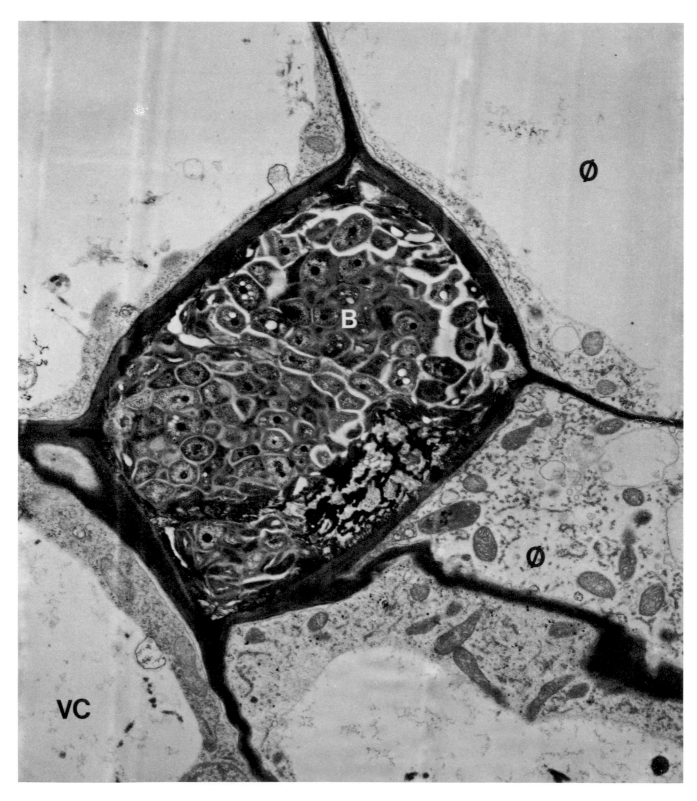

Fig. 69. Bacteria in the young cortex of chickpea. If pathogenic bacteria invade the root, there may be a strong host reaction. The cytoplasm of some cells may break down, (∅) and the host may begin to secrete materials (which stain intensely) into the intercellular space. B = bacterium, VC = vacuole. (×12,000)

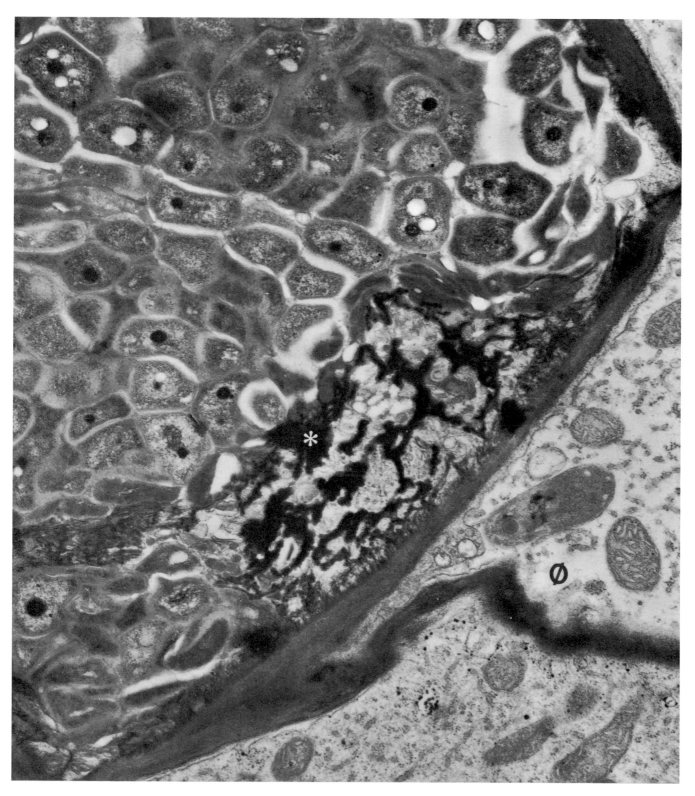

Fig. 70. Bacteria in the young cortex of chickpea. Detail of Fig. 69 shows the electron-dense material (*) accumulating near and apparently lysing associated bacteria in the intercellular space. The cytoplasm of cell (∅) is breaking down. (×43,000)

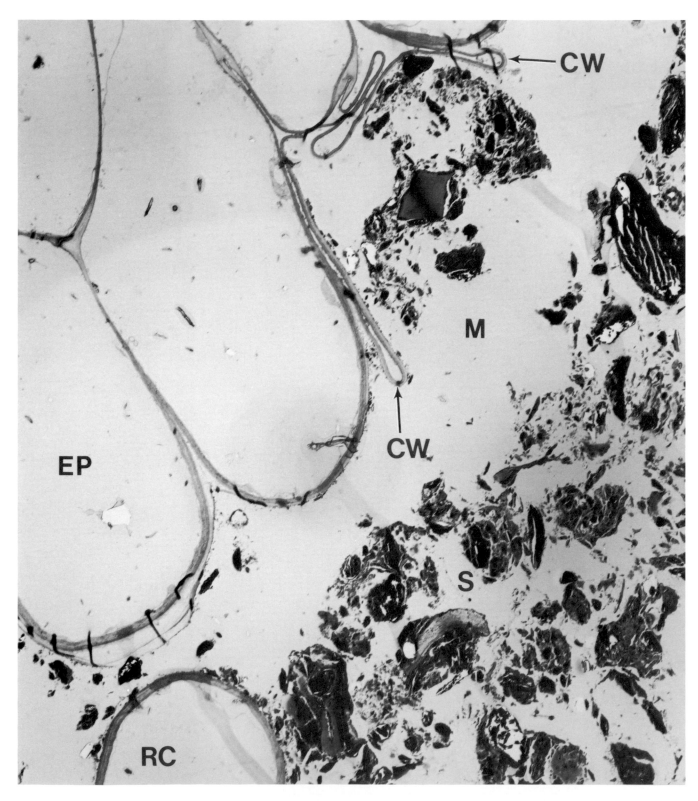

Fig. 71. Collapse of the epidermal cells of wheat. Autolysis of the epidermal cells (EP) is followed by their collapse, and minerals enter the space formerly occupied by the cell lumen. The radial walls (CW) remain, however, jutting out into the soil fabric. The space between the root and the clay fabric is occupied by mucilage (M), which is not stained in this preparation. RC = root cap cell (detached), S = soil. (×4,000)

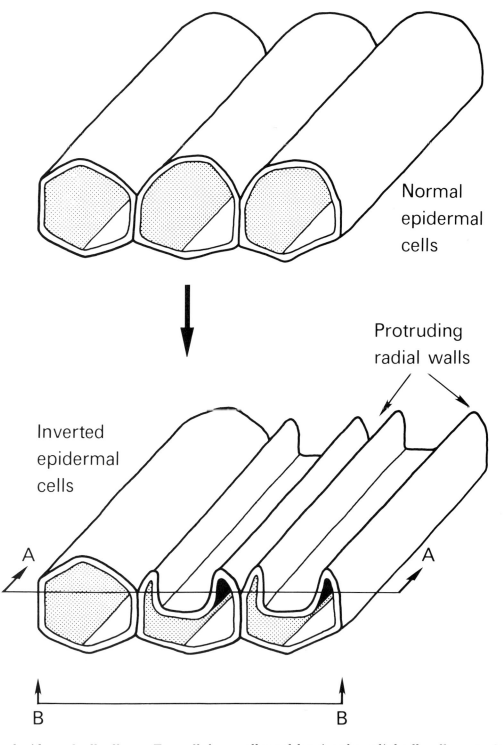

Normal
epidermal
cells

Protruding
radial walls

Inverted
epidermal
cells

A — A

B — B

Fig. 72. Stages of epidermal cell collapse. Two cells have collapsed, leaving the radial cell walls protruding into the soil. A and B indicate the planes of section in Figs. 75 and 76, respectively. Drawing by G. E. Rinder and R. M. Schuster.

Fig. 73. Collapse of the epidermal cells of wheat. The inversion of the epidermal cells is apparent in some SEM pictures. Note the projecting radial walls (arrows) and the clay particles occupying the space between. (×1,400)

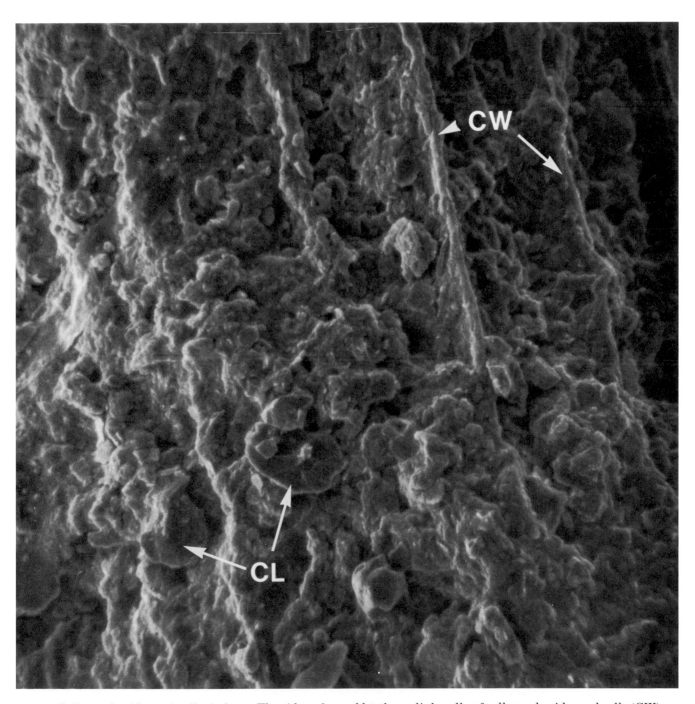

Fig. 74. Collapsed epidermal cell of wheat. The ridges formed by the radial walls of collapsed epidermal cells (CW) are particularly clear in this SEM. CL = clay. (×2,000)

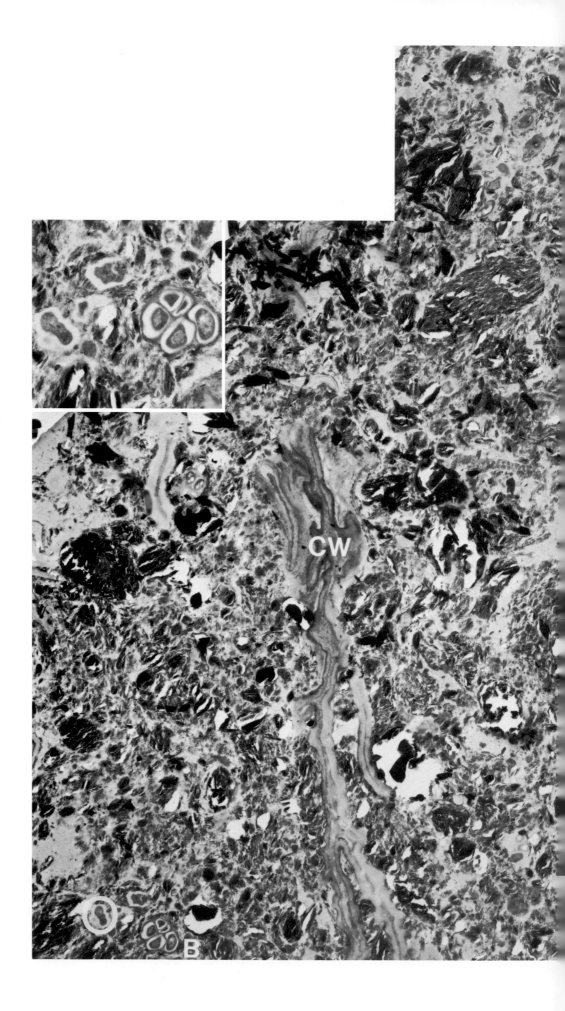

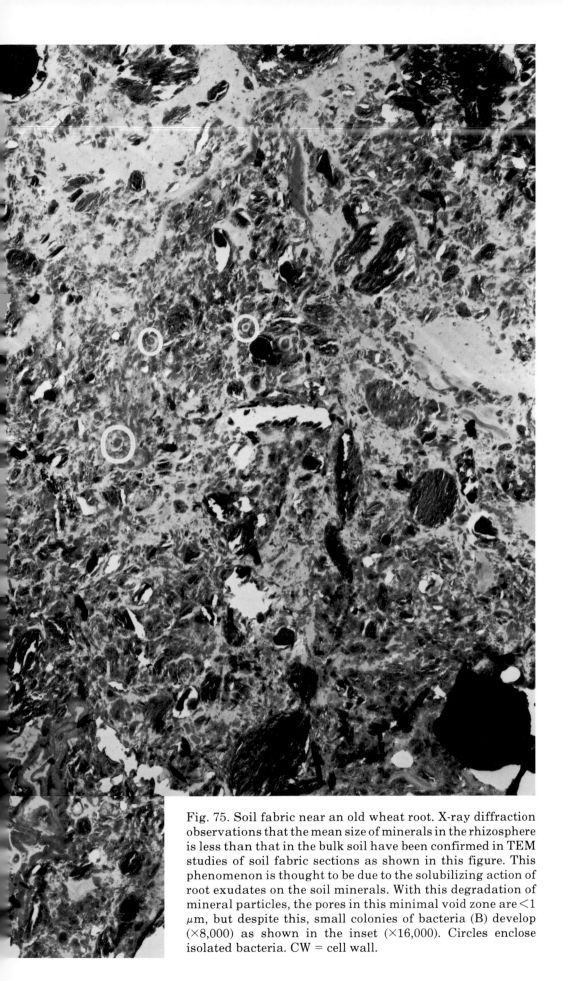

Fig. 75. Soil fabric near an old wheat root. X-ray diffraction observations that the mean size of minerals in the rhizosphere is less than that in the bulk soil have been confirmed in TEM studies of soil fabric sections as shown in this figure. This phenomenon is thought to be due to the solubilizing action of root exudates on the soil minerals. With this degradation of mineral particles, the pores in this minimal void zone are <1 μm, but despite this, small colonies of bacteria (B) develop (×8,000) as shown in the inset (×16,000). Circles enclose isolated bacteria. CW = cell wall.

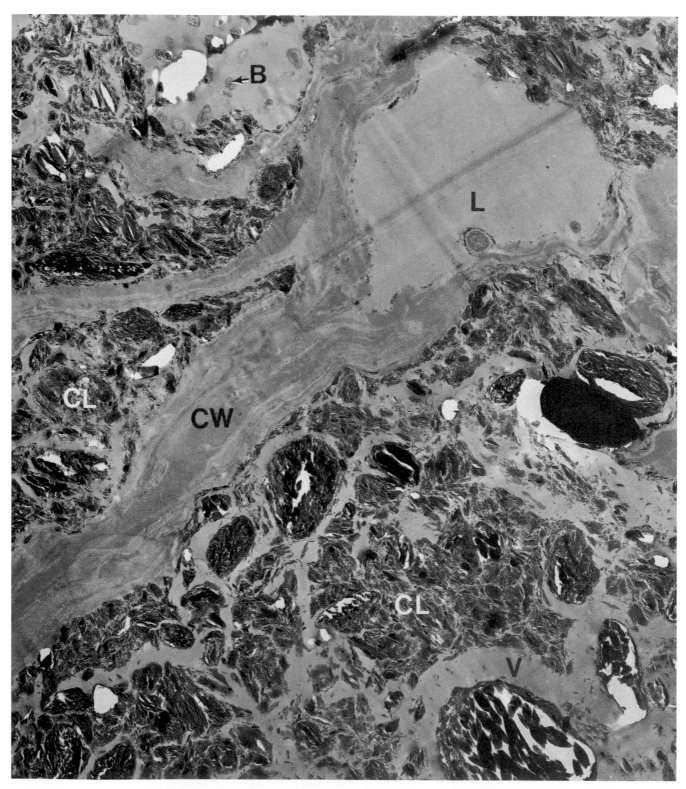

Fig. 76. Soil near a wheat root surface. Detail of the root surface shows the protruding radial cell walls (CW) and the lumena (L) of collapsed epidermal cells, one of which (CL) is partially occupied by clay minerals. A colony of bacteria (B) lies near one cell. V = void. (×4,000)

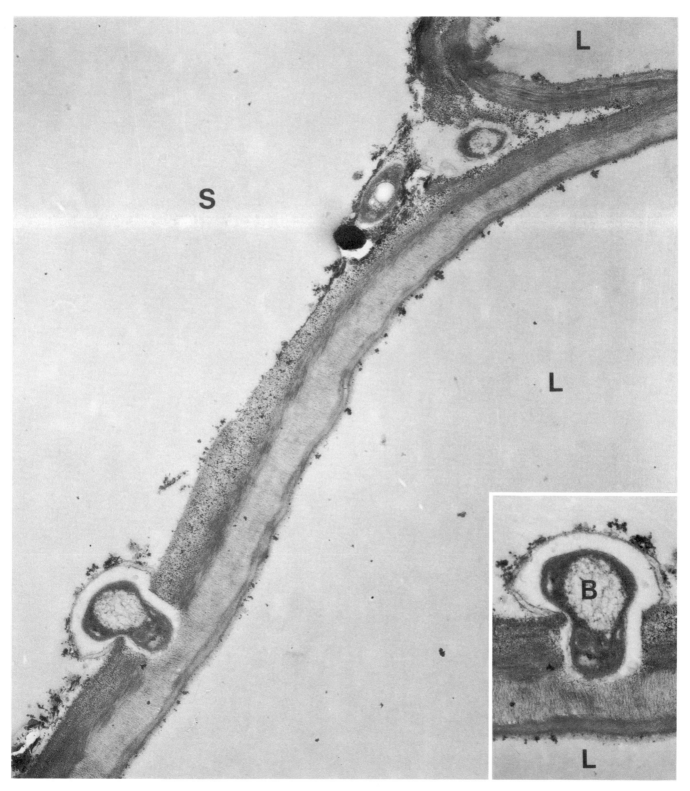

Fig. 77. Invasion of the cortical cells of wheat. L = epidermal cell lumen, S = soil. (×26,000) In the inset, a microorganism (B) has penetrated the outer primary wall and is attacking the more resistant inner lignified secondary wall. Other bacteria are lysing the gel in the intercellular space between the epidermal cells. (×34,000) Reprinted, by permission, from R. C. Foster and A. D. Rovira, Ultrastructure of wheat rhizosphere, New Phytol. 76:343-352, Plate 4, Fig. 7, ©1976.

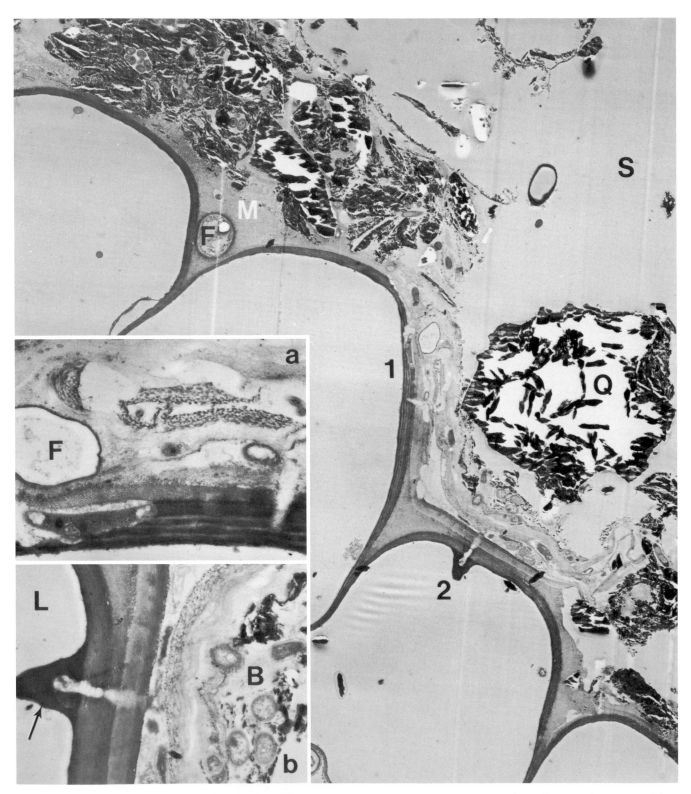

Fig. 78. Invasion of cortical cells of *Paspalum*. Pockets of mucilage (M) may survive where the root is protected by a dense soil fabric, but microbial lysis of the gel is rapid and almost complete where the soil fabric is open. (×3,600) Inset **a,** an enlargement of site 1, shows microbial decay of both the outer and inner multilamellate cell wall layers. (×17,000) Inset **b,** an enlargement of site 2, shows the lignituber formed by the host in response to microbial attack (arrow). (×12,000) B = bacterium, F = fungal hypha, L = cell lumen, Q = quartz, S = soil. Reprinted, by permission, from R. C. Foster, The ultrastructure and histochemistry of the rhizosphere, New Phytol. 89:263-273, Plate 4, Fig. 1, ©1981.

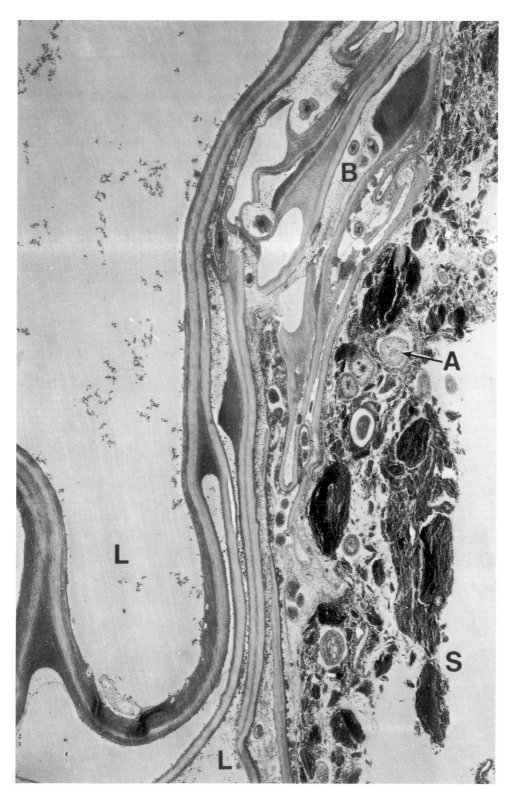

Fig. 79. Collapse of the outer cortex of wheat. Microbial lysis of the cell wall, combined with root pressure, causes the collapse of the outer layers of the cortex so that their lumena (L) are obliterated. The dense staining of some of the cell walls suggests that they are suberized and/or lignified. The bacteria within the root are a different shape and size than those in the rhizosphere. A = actinomycete, B = bacterium, S = soil. (×10,000) Reprinted, by permission, from R. C. Foster and A. D. Rovira, Ultrastructure of wheat rhizosphere, New Phytol. 76:343-352, Plate 1, ©1976.

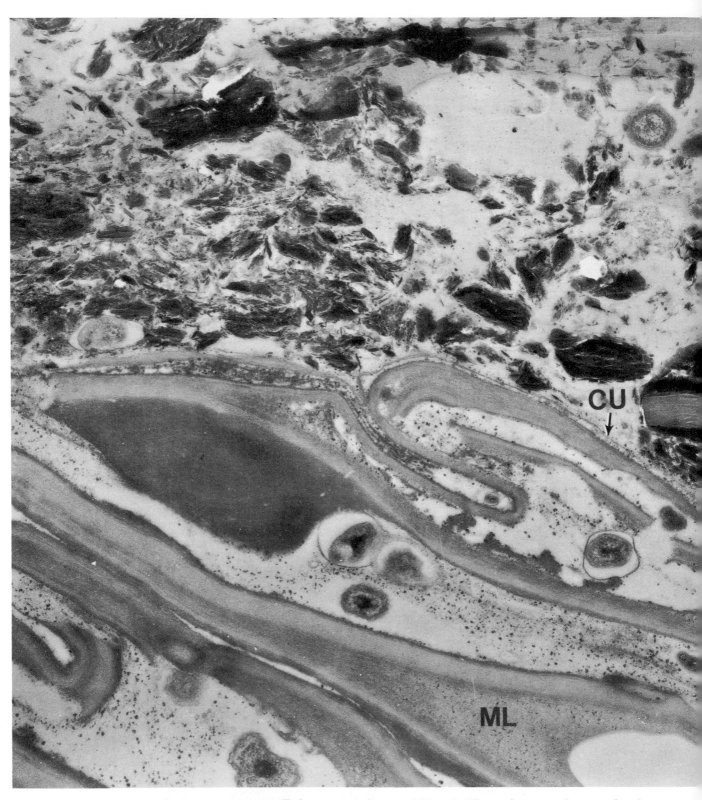

Fig. 80. Details of collapsed wheat root cortex. Enlargement of part of Fig. 79. These electron micrographs show a wide range of microorganisms both within and at the surface of the root. Some organisms lie free in the clay fabric. Others are enclosed in capsule material. Despite the removal of the mucilage, the fine cuticle (CU) still encloses the root. In addition to microbial lysis, there has apparently been mechanical breaking of the cell walls. The lignified middle lamella (ML) substances have not yet been attacked. Many of the clay (CL) particles have been reoriented by root pressure so that they lie with their narrowest dimension parallel to the root surface. (×30,000) Reprinted, by permission, from R. C. Foster and A. D. Rovira, Ultrastructure of wheat rhizosphere, New Phytol. 76:343-352, Plate 1, ©1976.

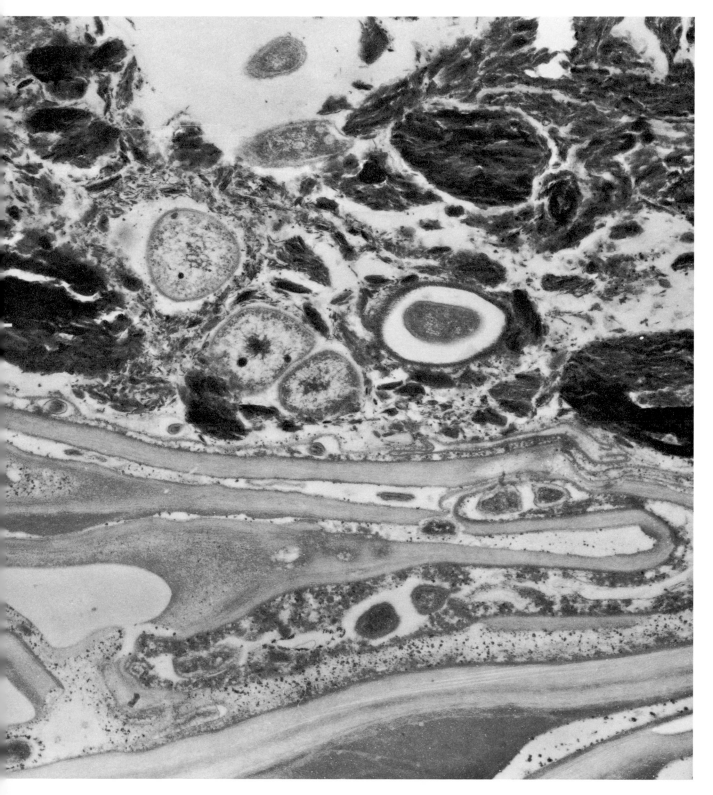

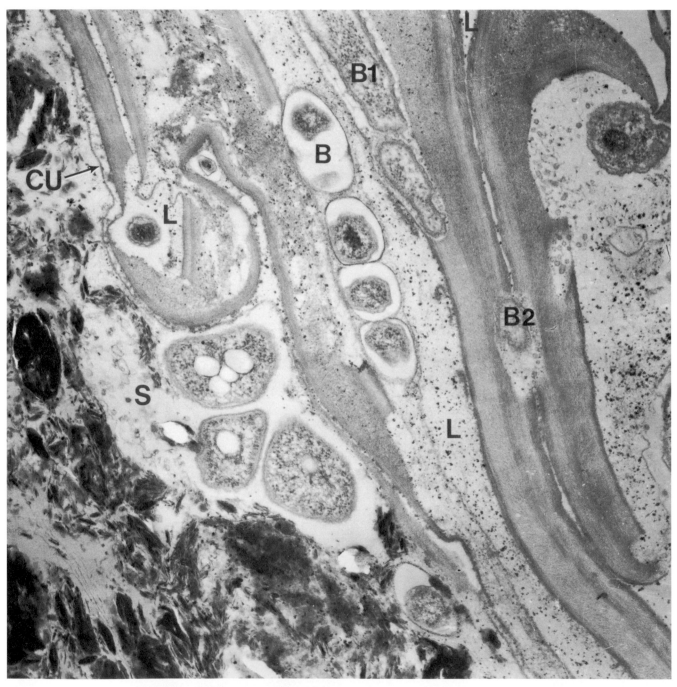

Fig. 81. Details of cortical cell lysis in wheat. Cell wall lysis may be accomplished by symbiosis between several types of bacteria (B). Some (B1) attack only the cellulose-rich layers of the cell wall, leaving the terminal and middle lamellae on either side unlysed. Others (B2) are able to lyse the lignified layers. Again the root is bound by the cuticle (CU). L = cell lumen, S = soil. (×33,000) Reprinted, by permission, from R. C. Foster and A. D. Rovira, Ultrastructure of wheat rhizosphere, New Phytol. 76:343-352, Plate 2, Fig. 3, ©1976.

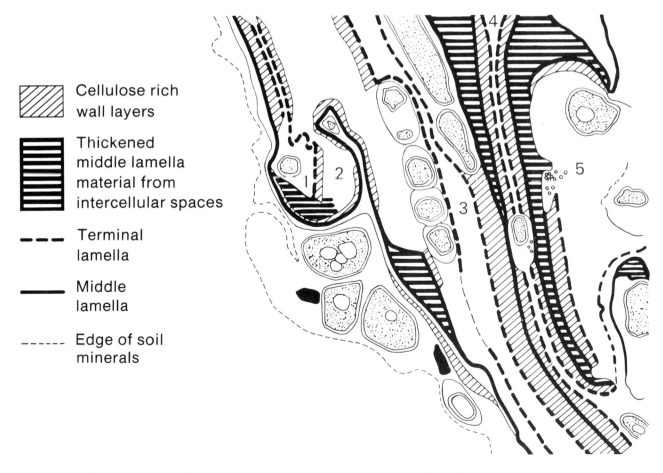

Cellulose rich
wall layers

Thickened
middle lamella
material from
intercellular spaces

Terminal
lamella

Middle
lamella

Edge of soil
minerals

Fig. 82. Cortical cell collapse. Numbers 1–5 are the lumena of the outer collapsed cortical cells, showing progressive microbial breakdown of the cell wall material. The innermost cell, cell 5, is least decayed; cell 1 has largely been reduced to the resistant terminal and middle lamella layers. In cells 1 and 4, collapse has been so severe that the terminal lamellae from opposite sides of the cell are almost touching and the lumen has almost disappeared. (Adapted from Fig. 1, New Phytol. 76:343-352, 1976)

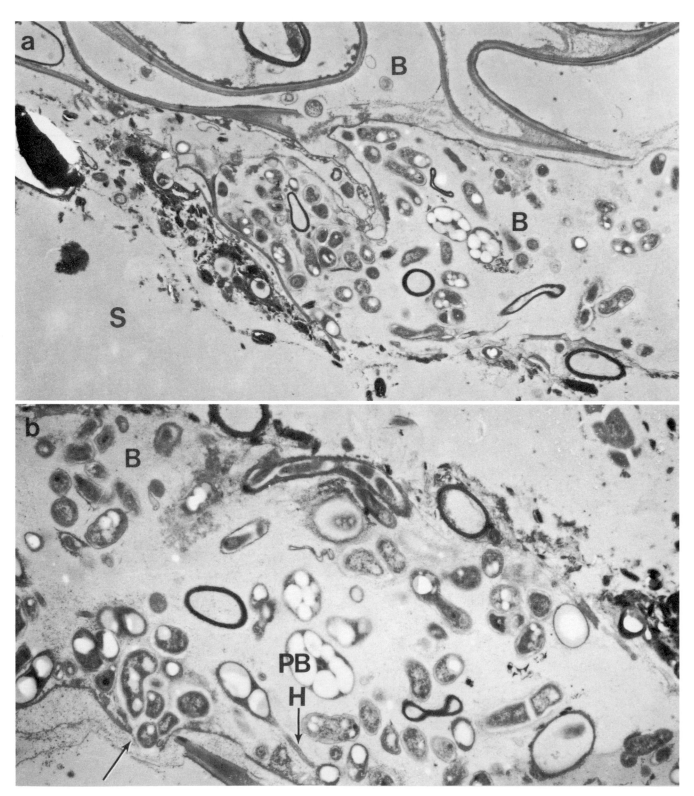

Fig. 83. Colonization of the epidermal and cortical cells of wheat. **a,** The lumen of the cortical cells eventually becomes filled with a mixed population of bacteria (B) and actinomycetes, some of which form thick-walled spores. S = soil. (×13,000) **b,** The lumen of this epidermal cell has become filled with a range of microorganisms including *Hyphomicrobium* (H) with its characteristic shape. Many cells are filled with polyhydroxybutyrate (PB) and appear as white areas inside cells. A group of cells have broken down the internal tangential wall and are about to enter the cell below (arrow). (×21,000) Reprinted, by permission, from R. C. Foster and A. D. Rovira, Ultrastructure of wheat rhizosphere, New Phytol. 76:343-352, Plate 6, Fig. 11, ©1976.

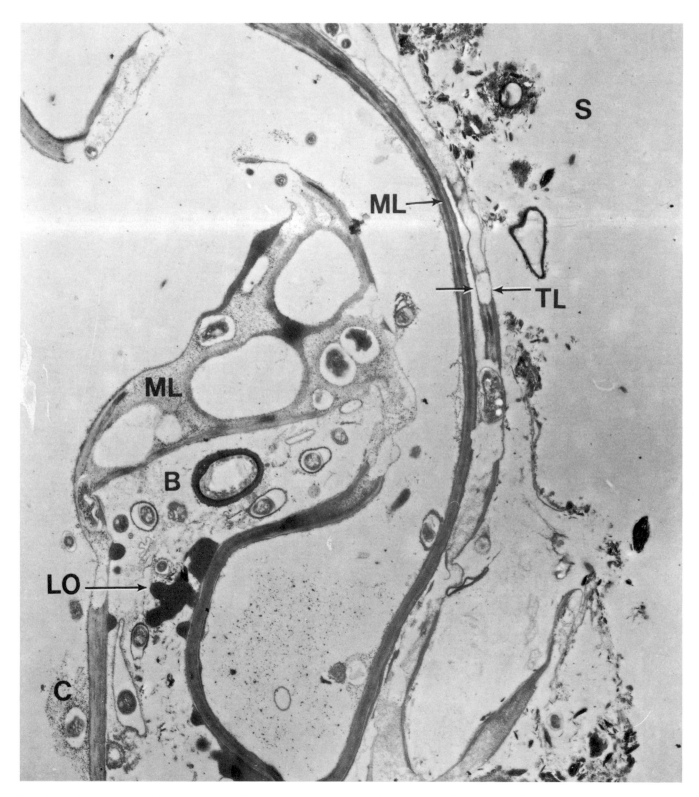

Fig. 84. Colonization of the cortex in wheat. A mixed microbial population has broken down the outer cortical cells. Lignified terminal lamellae (TL) still remain intact at localized points, but other parts are being hydrolyzed by bacteria (arrows). B = bacterium, C = capsule material, LO = cell wall lobes, M = middle lamella, S = soil. (×13,000)

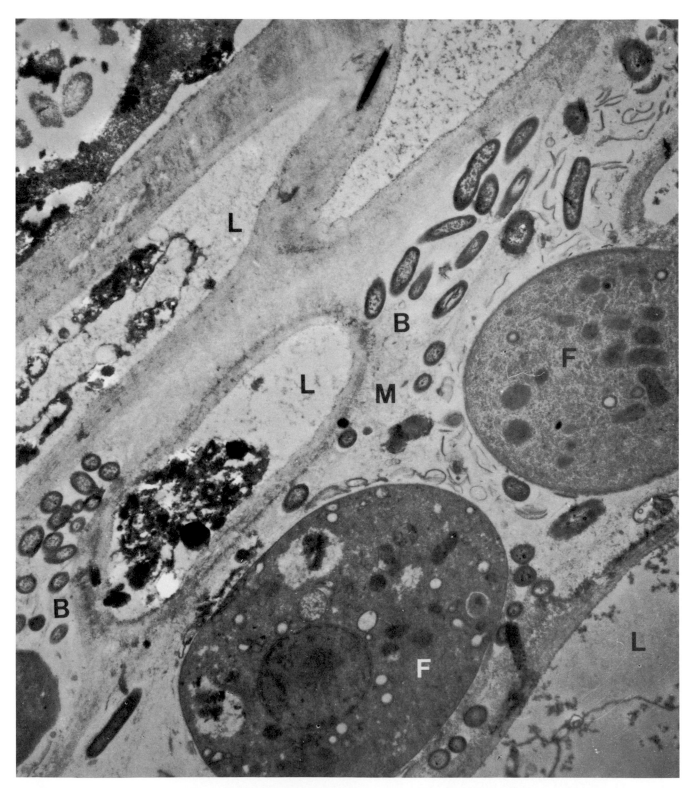

Fig. 85. Secondary invaders in wheat infected by cereal cyst nematode. This section through a wheat root invaded by *Heterodera avenae* shows that the thick mucilage induced by nematode attack has become infected by a variety of secondary invaders including fungi (F) and bacteria (B). This increase in polysaccharides at the point of infection by cereal cyst nematode may play a role in the increased damage by *Rhizoctonia* in the presence of cereal cyst nematodes. L = cortical cell lumen, M = mucigel, S = soil. (×30,000)

Mycelial Strands
and Mycorrhizas

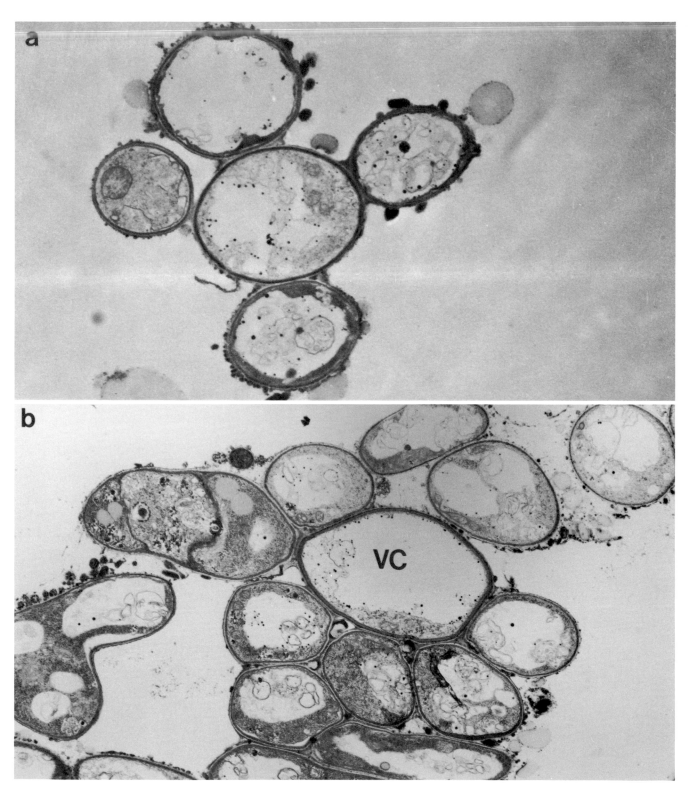

Fig. 86. Young mycelial strand from a pine mycorrhiza. **a,** Mycelial strands of pines arise as a collection of hyphae that surround a central, often larger hypha near the surface of the mantle of a mycorrhizal root. The peripheral hyphae are filled with cytoplasm and have small vacuoles. (×12,500) Reprinted, by permission, from R. C. Foster, Mycelial strands of *Pinus radiata* D. Don: Ultrastructure and histochemistry. New Phytol. 88:705-712, Plate 1, Fig. 1, ©1981. **b,** Mycorrhizas are formed by both Basidiomycetes and Ascomycetes: the former form characteristic dolipore septa unique to the group. The septum has a central pore enclosed in bracketlike parenthosomes, and the whole is enclosed on either side by a perforated elaboration of the endoplasmic reticulum. Dolipore septa can be seen in both longitudinal and transverse section. The cytoplasm of the central hypha (VC) is beginning to become disorganized. (×9,000)

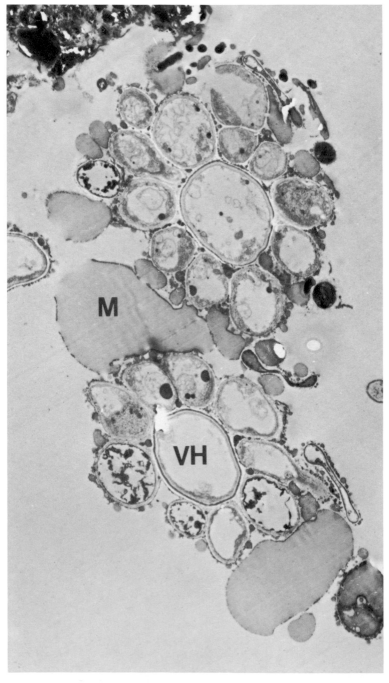

Fig. 87. Aggregation of young mycelial strands of pines. Adjacent aggregations of hyphae become associated laterally to form more complex strands. By this stage, the central hyphae are almost devoid of cytoplasm. Gel may accumulate near the strand (M). VH = vessel hypha. (×6,000)

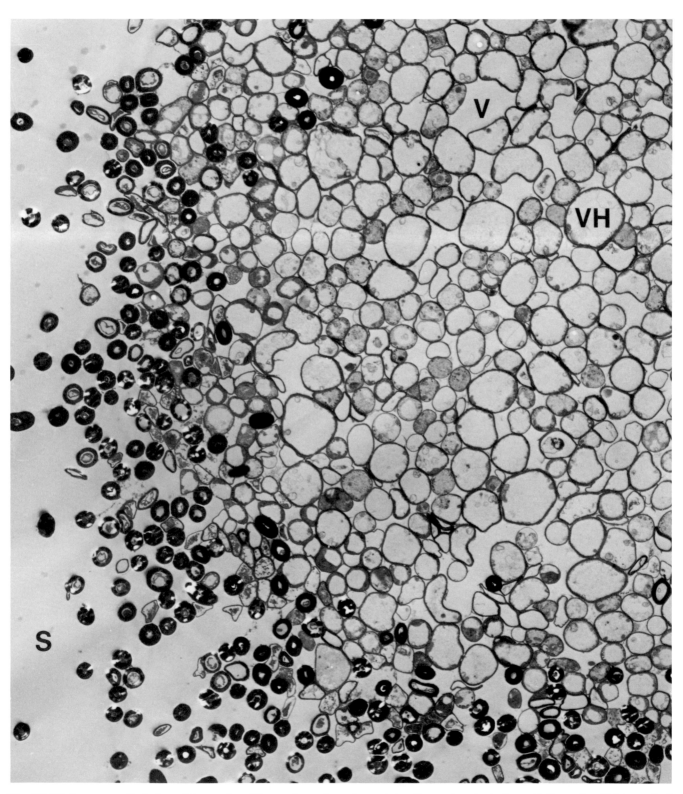

Fig. 88. Mature mycelial strand from pine. The mature strand consists of an outer rind of small diameter, thick-walled, widely spaced, dead hyphae that stain intensely with heavy metals. The core consists of an open network of thin-walled vessel (VH) and living hyphae. S = soil, V = void. (×3,300) Reprinted, by permission, from R. C. Foster, Mycelial strands of *Pinus radiata* D. Don: Ultrastructure and histochemistry. New Phytol. 88:705-712, Plate 3, Fig. 2, ©1981.

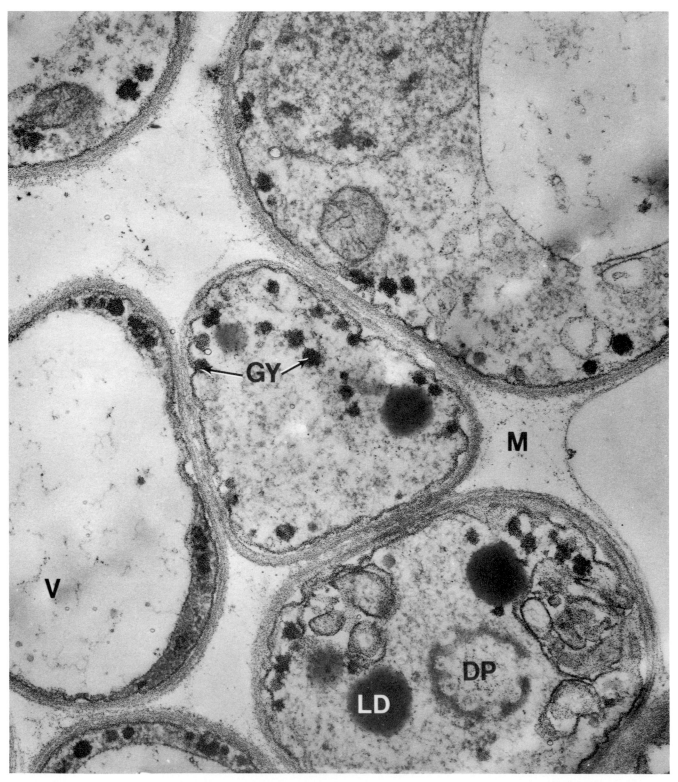

Fig. 89. Detail of strand hyphae. The core hyphae contain granules of glycogen (GY) and oil droplets (LD). The individual cells are enclosed in a rather tenuous gel (M). Note the tangential section of the dolipore septum (DP). V = vacuole. (Periodic acid-thiosemicarbazide silver proteinate stain, ×55,000)

Fig. 90. Outer rhizosphere of an ectotrophic mycorrhiza of pine. The rhizospheres of ectotrophic mycorrhizas support an extensive microbial population (B). But most of them are confined to a zone immediately outside the mantle (MA). The soil (S) contains humified organic masses (HM) as well as minerals. The outer layers of the mantle are often devoid of cytoplasm, and the cells are extended tangentially. Within the mantle are the tannin-filled cells (PT) of the host epidermis. (×8,000) Reprinted, by permission, from R. C. Foster and G. C. Marks, Observations on the mycorrhizas of forest trees. II. The rhizosphere of *Pinus radiata* D. Don, Aust. J. Biol. Sci. 20:915-926, Plate 4, ©1967.

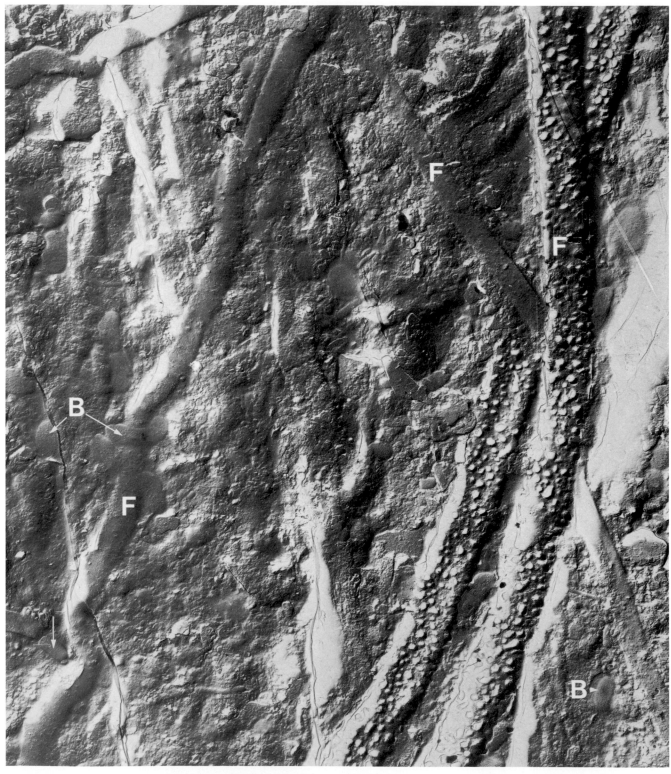

Fig. 91. Surface ectotrophic mycorrhiza. Some mycorrhizal hyphae are coated with oxalate crystals. In some soils, 50% of the weight of the mycorrhizal mat consists of oxalate, which is secreted by the mycorrhizal fungus and may chelate ions from clay particles and accelerate the weathering of minerals. Note bacteria (B) and fine clay particles attached to the mycorrhizal surface. (Platinum-palladium shadowed replica). F = fungal hypha. (×6,000)

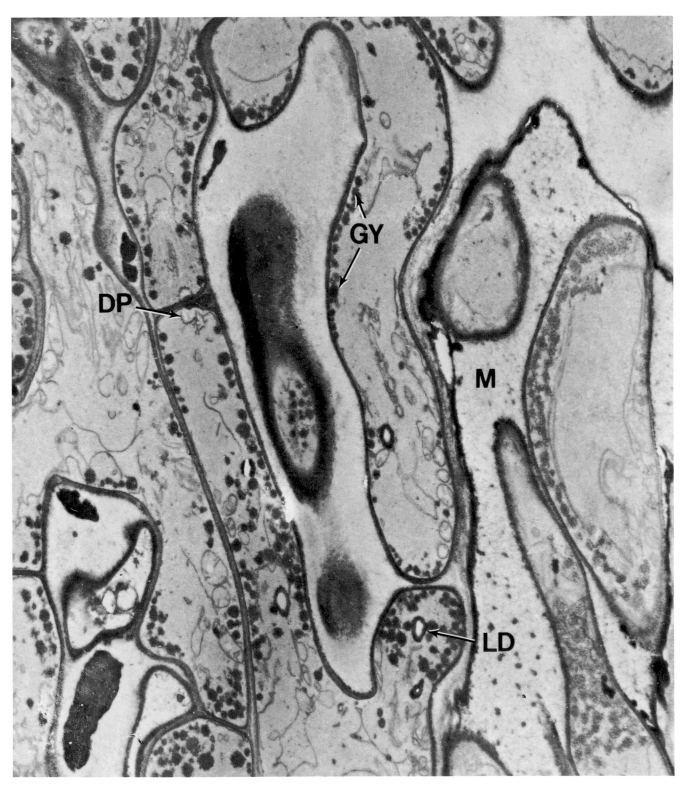

Fig. 92. Mantle of a pine mycorrhiza. The mantle hyphae are filled with glycogen granules (GY) and occasional oil droplets (LD). The cells may be immersed in a granular or fibrous gel (M). DP = dolipore septum. (×11,000) Reprinted, by permission, from R. C. Foster and G. C. Marks, The fine structure of the mycorrhizas of *Pinus radiata* D. Don, Aust. J. Biol. Sci. 19:1027-1038, Plate 2, Fig. 1, ©1966.

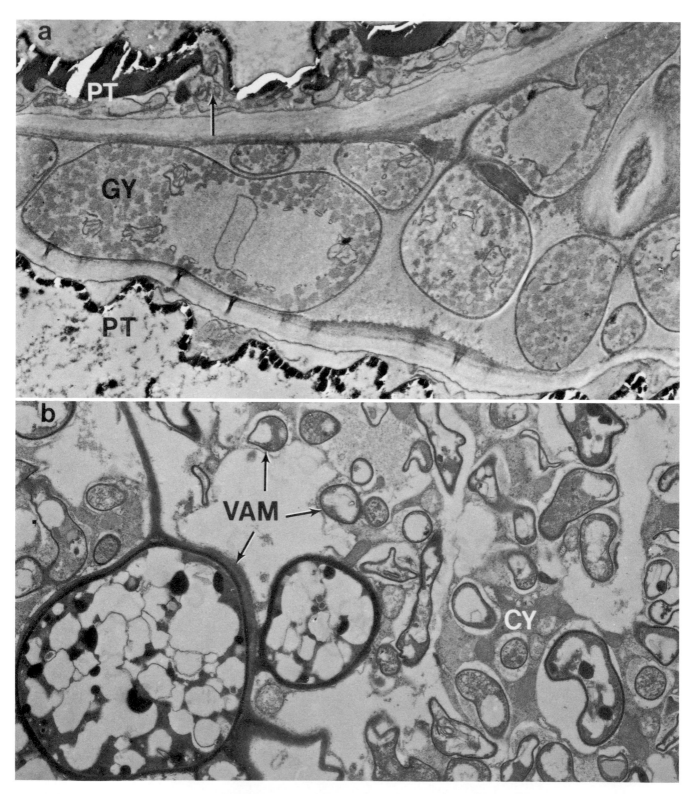

Fig. 93. Mycorrhizas in root cortex of (a) pine and (b) clover. **a,** Ectotrophic mycorrhizas. The hyphae of the ectotrophic mycorrhizas penetrate and fill the intercellular spaces of the outer cortex to form the Hartig net. The hyphae are filled with glycogen granules (GY) and occasional oil droplets. The host cells secrete tannin (PT) into the vacuoles, and it may be significant that the amyloplasts (arrow) of the adjacent host cells are generally void of starch. (×14,000) **b,** Endotrophic mycorrhizas. The hyphae of vesicular-arbuscular mycorrhizal (VAM) fungi penetrate the cell walls of the cortical cells of the host, enter the cell lumen, and occupy the intercellular spaces. Here they proliferate to form treelike branching systems (arbuscules) within the host cytoplasm, thus increasing the area of contact between the host cytoplasm (CY) and fungus through which nutrients are transferred between the symbionts. (×10,000)

Soilborne Root
Disease Organisms
and Their Biological Control

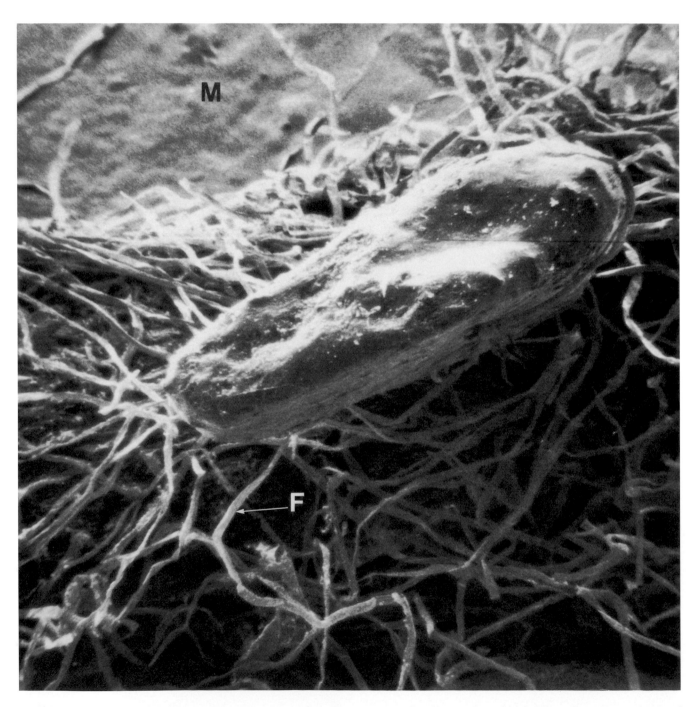

Fig. 94. Growth on root of *Gaeumannomyces graminis* from propagule of ground oat kernel inoculum. Experimental infection with take-all (*G. graminis* var. *tritici*) can be achieved by mixing soil with ground oat seeds on which the fungus has grown. F = fungus, M = mucilage. (×250)

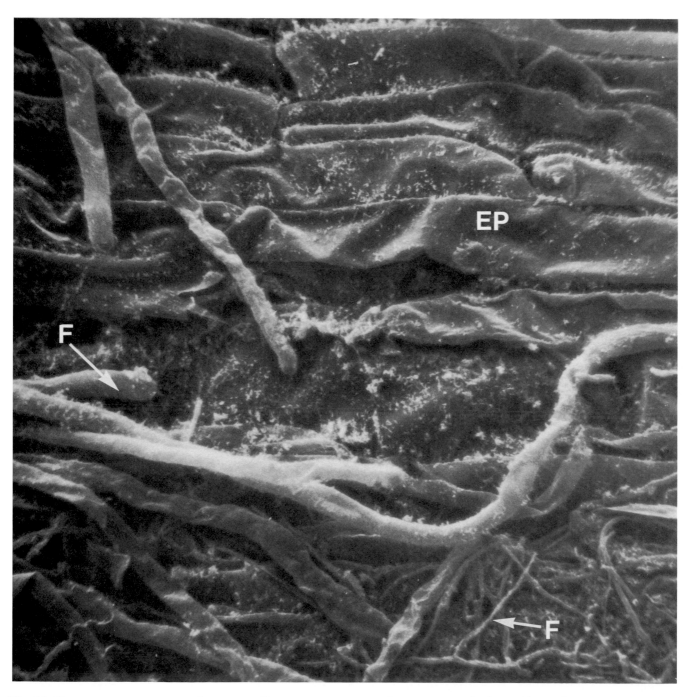

Fig. 95. *Gaeumannomyces* on a wheat root. *Gaeumannomyces* produces a network of mycelia on the root. Two types of hyphae (F) are produced: broad dark runner hyphae grow in the rhizosphere parallel to the root, and narrow hyaline, infective hyphae grow on the root and penetrate the cortex. EP = epidermal cell. (×4,000) Reprinted, by permission, from A. D. Rovira and R. Campbell, A scanning electron microscope study of the interactions between microorganisms and *Gaeumannomyces graminis* (syn. *Ophiobolus graminis*) on wheat roots, Microb. Ecol. 2:177-185, Fig. 1, ©1975, Springer-Verlag, Inc., New York.

Fig. 96. Entry of hypha into wheat root. Here a broad runner hypha (F) has given rise to two narrow infective hyphae, one of which is penetrating the epidermal cell wall. The remains of a lysed hypha (arrow) occurs on the root surface below the branched hypha. The branched hypha itself is not healthy, as indicated by the rough surface and the dark patches. (×28,000) Reprinted, by permission, from A. D. Rovira and R. Campbell, A scanning electron microscope study of the interactions between microorganisms and *Gaeumannomyces graminis* (syn. *Ophiobolus graminis*) on wheat roots, Microb. Ecol. 2:177-185, Fig. 16, ©1975, Springer-Verlag, Inc., New York.

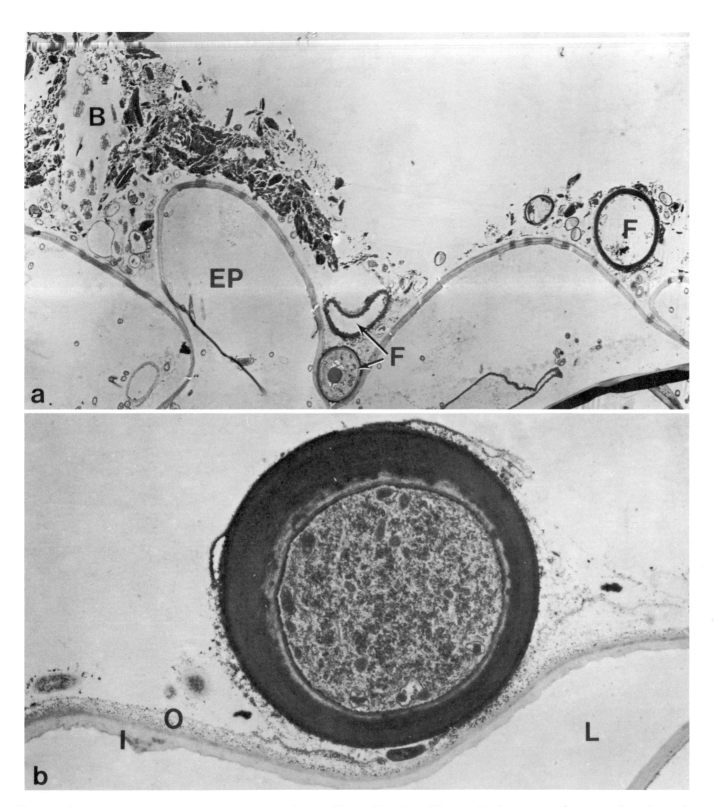

Fig. 97. *Gaeumannomyces* on a wheat root surface. **a,** Fungal hyphae (F) grow in the grooves between the cells and along the surface of an experimentally infected wheat plant. The epidermal cells (EP) have become infected by bacteria (B), and numerous bacteria occur in the rhizoplane. (×4,400) **b,** The runner hyphae have a thick cell wall, which in this specimen is just completing melanization. The hypha is embedded in a fine granular mucigel and a bacterium lies in the mucigel between the hypha and the host cell. Such melanized hyphae are more resistant to microbial attack than the finer hyaline hyphae. I = inner layer of epidermal cell wall, L = epidermal cell lumen, O = outer layer of epidermal cell wall. (×16,000)

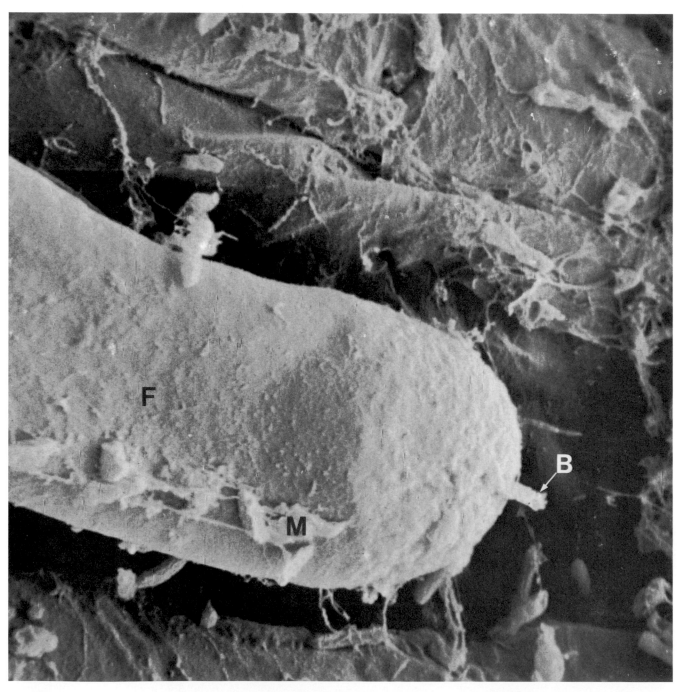

Fig. 98. Tip of *Gaeumannomyces* hypha on wheat. Fungal hyphae (F) grow only at the apex, which is covered with a fine granular material. Fragments of root mucilage (M) adhere to the hypha, and bacteria (B) are attached end on. (×4,000)

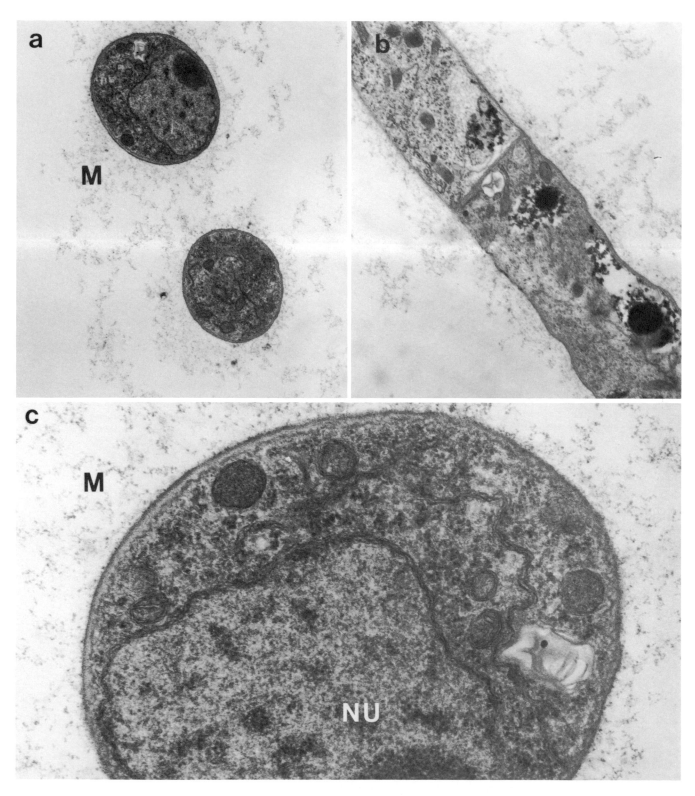

Fig. 99. *Gaeumannomyces* hyphae in cortex of wheat. Having entered the root, hyphae grow through the cortex and proliferate in the stele where they are seen in transverse (**a, c**) and longitudinal sections (**b**). Inside the root, the hyphae are surrounded by a tenuous mucilage (M). The hyphae have a thin cell wall and the usual complement of cytoplasmic organelles: nucleus (NU), mitochondria, endoplasmic reticulum, and ribosomes. The vacuoles are small and often contain electron-dense material. (a, ×8,500; b, ×8,500; c, ×51,000)

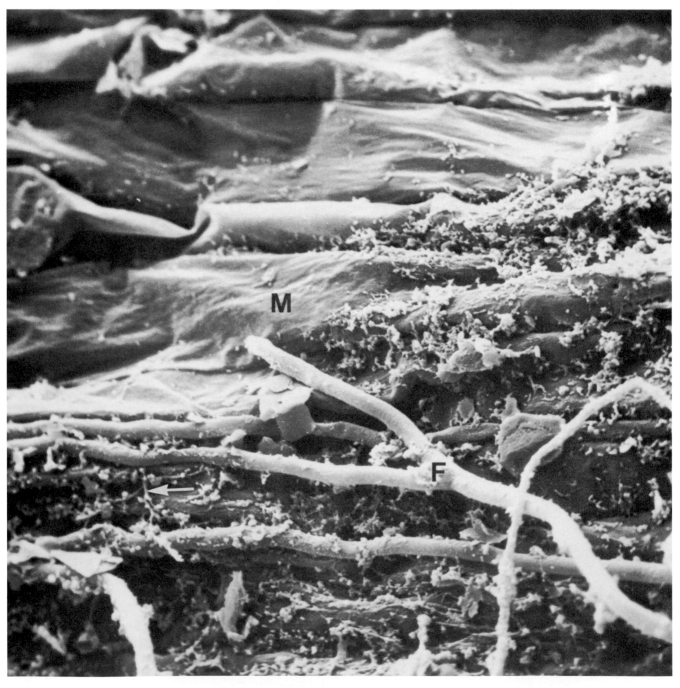

Fig. 100. Edge of a *Gaeumannomyces* lesion in wheat. Entry of the *Gaeumannomyces* hyphae induces the release of nutrients from the root cells, and large populations of bacteria build up in the mucigel (M) on the epidermal cells. At first these bacteria are covered by the mucigel, but this is later removed by microbial lysis. Mucigel lysis results from the combined action of the fungal hyphae (F), bacteria, and actinomycetes (arrow). (×4,000) SEM courtesy of R. Campbell, University of Bristol

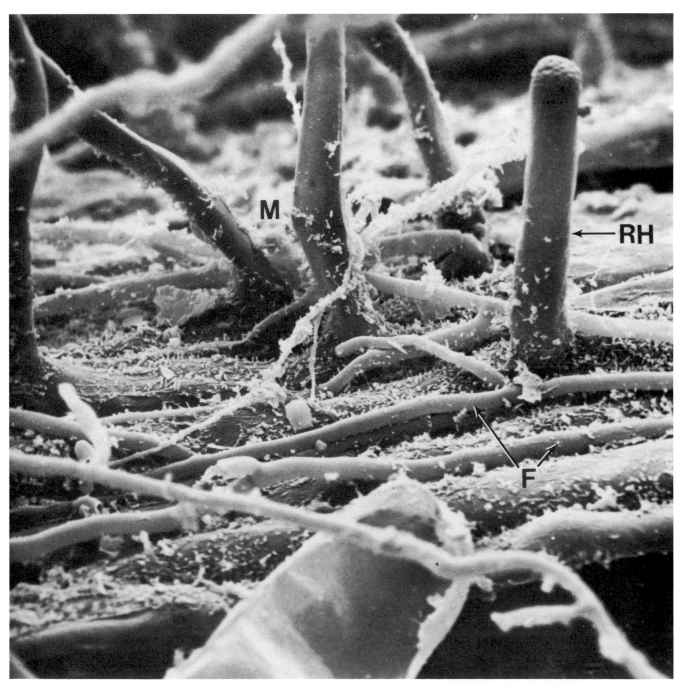

Fig. 101. Lesion in the root hair zone of wheat. The mucigel (M) in the root hair (RH) zone has disappeared following infections with *Gaeumannomyces*, and there is a dense colonization by rod-shaped bacteria. F = fungus. (×4,000) SEM courtesy of R. Campbell, University of Bristol

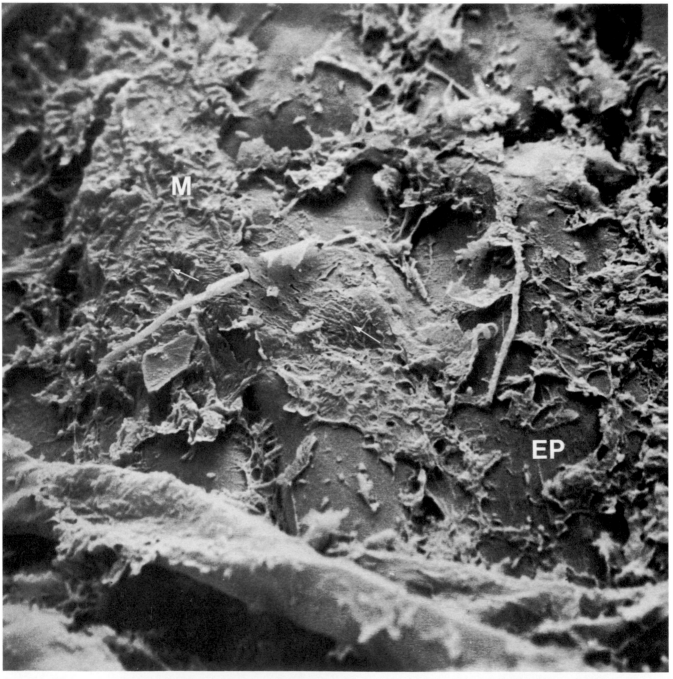

Fig. 102. Old lesion caused by *Gaeumannomyces* affecting wheat. Eventually the mucigel (M) is reduced to tattered remnants that are heavily colonized by bacteria (arrows). EP = epidermal cell. (×1,000) Reprinted, by permission, from A. D. Rovira and R. Campbell, A scanning electron microscope study of the interactions between microorganisms and *Gaeumannomyces graminis* (syn. *Ophiobolus graminis*) on wheat roots, Microb. Ecol. 2:177-185, Fig. 8, ©1975, Springer-Verlag, Inc., New York.

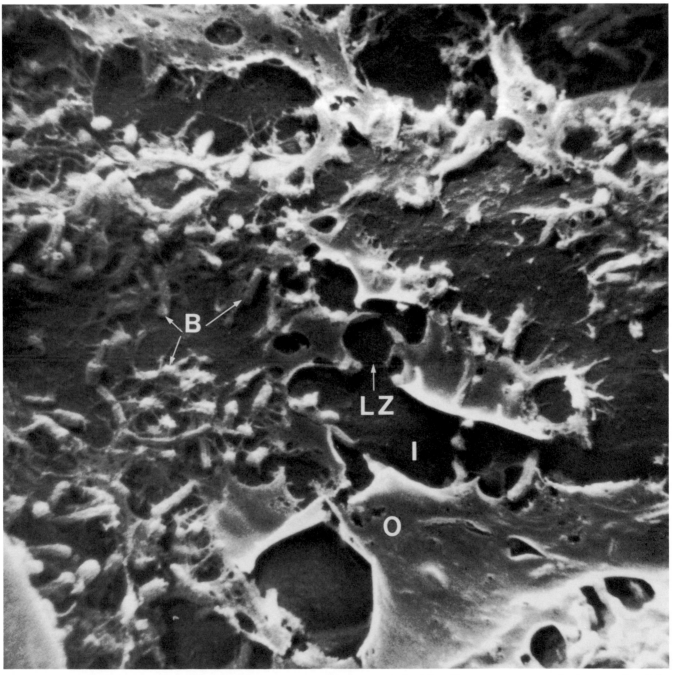

Fig. 103. Remnants of mucigel and rod-shaped bacteria (B), associated with a *Gaeumannomyces* lesion on wheat. I = inner layer of epidermal cell wall, LZ = lysis zone, O = outer layer of epidermal cell wall. (×6,000) Reprinted, by permission, from A. D. Rovira and R. Campbell, Scanning electron microscopy of microorganisms on the roots of wheat, Microb. Ecol. 1:15-23, Fig. 4c, ©1974, Springer-Verlag, New York.

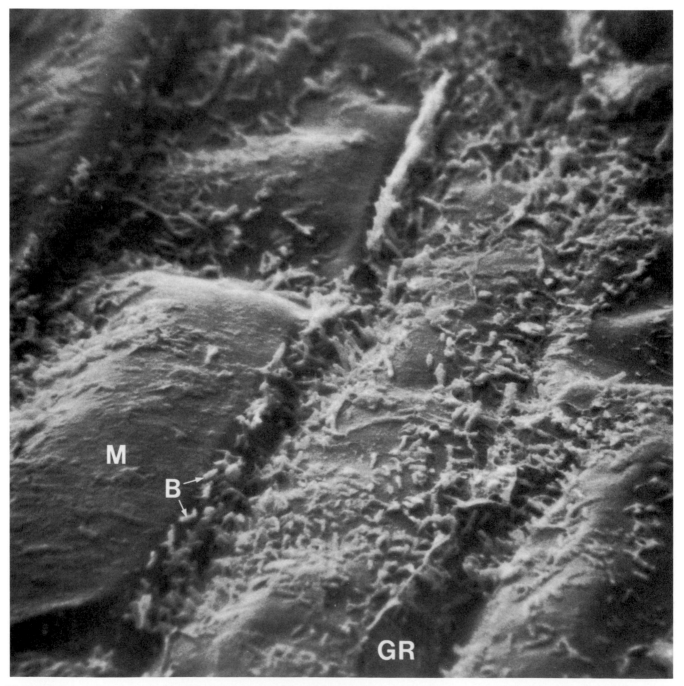

Fig. 104. Bacteria associated with lesion caused by *Gaeumannomyces* on wheat. Some root cells are more heavily colonized by bacteria than their neighbors. Here the mucigel (M) is being removed from the central cell, but the mucilage on neighboring cells is still intact. Bacteria (B) are spreading across the grooves (GR) onto these neighboring cells. (×2,800)

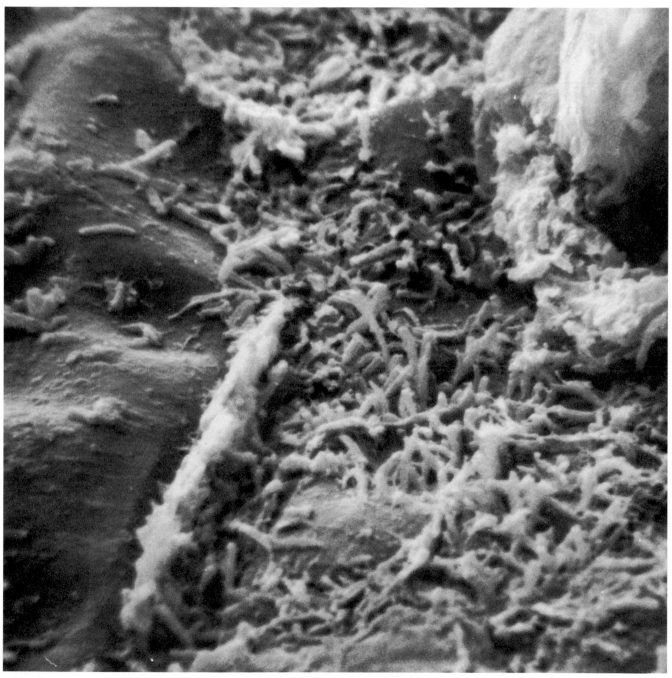

Fig. 105. The surface of a lesion caused by *Gaeumannomyces* in wheat. In diseased roots, the rod-shaped bacteria are densely packed in the mucigel; individual host cells seem to be colonized by bacteria of the same type, and this phenomenon of bacterial proliferation on roots infected with *Gaeumannomyces* is thought to be the precursor to take-all decline. (×5,600)

Fig. 106. TEM of a replica of a surface of a wheat root infected with *Gaeumannomyces*. Removal of the mucigel by the bacteria and fungi exposes the cellulose microfibrils of the host cell wall. (×33,000)

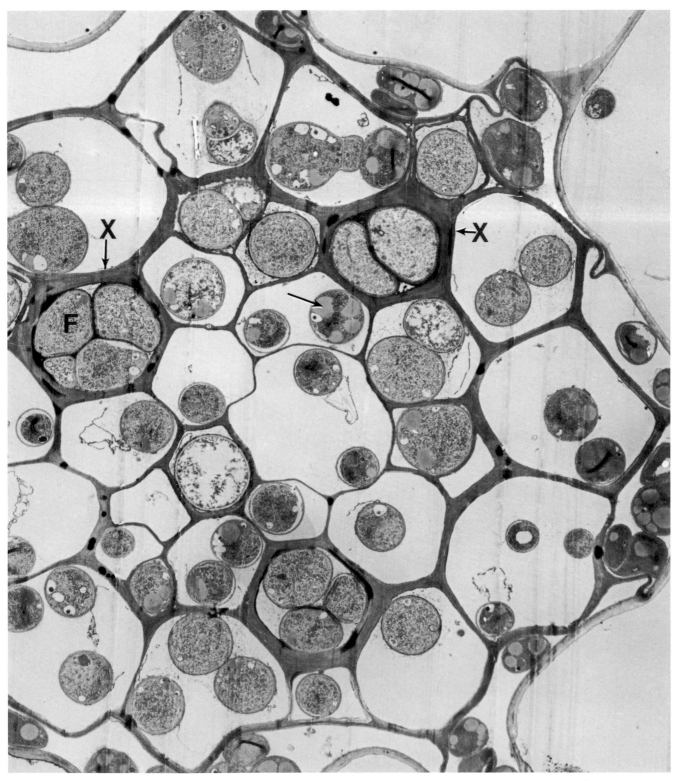

Fig. 107. *Gaeumannomyces* in the stele of wheat. In this root from a heavily infected seedling with classic symptoms of take-all, the *Gaeumannomyces* hyphae have colonized the vascular tissues of the stele. Most of the lumen of the tracheids (X) has been occupied by hyphae, preventing the movement of water from the roots to the leaves. The hyphae are beginning to store (arrow) lipid. F = fungal hypha. (×3,900)

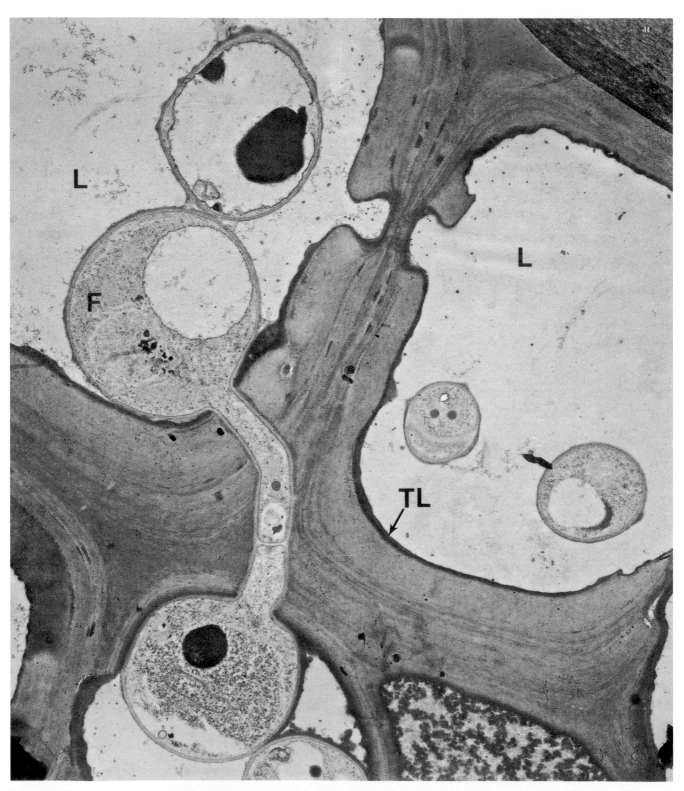

Fig. 108. *Gaeumannomyces* in the stele of a root of a wheat plant showing premature ripening or "hay die." Here the hyphae (F) of *Gaeumannomyces* have colonized the thick-walled, fully differentiated tracheids of a mature wheat plant. In addition to growing axially along the vascular tissues, the fungus forms narrow lateral infection hyphae that penetrate the cell wall of the host so that the fungus is able to move laterally into new cells. When the infection hypha enters the new host cell, it swells to form a hypha of normal dimensions. A septum then forms in the infection hypha. L = lumen of tracheid, TL = terminal lamella. (×11,500)

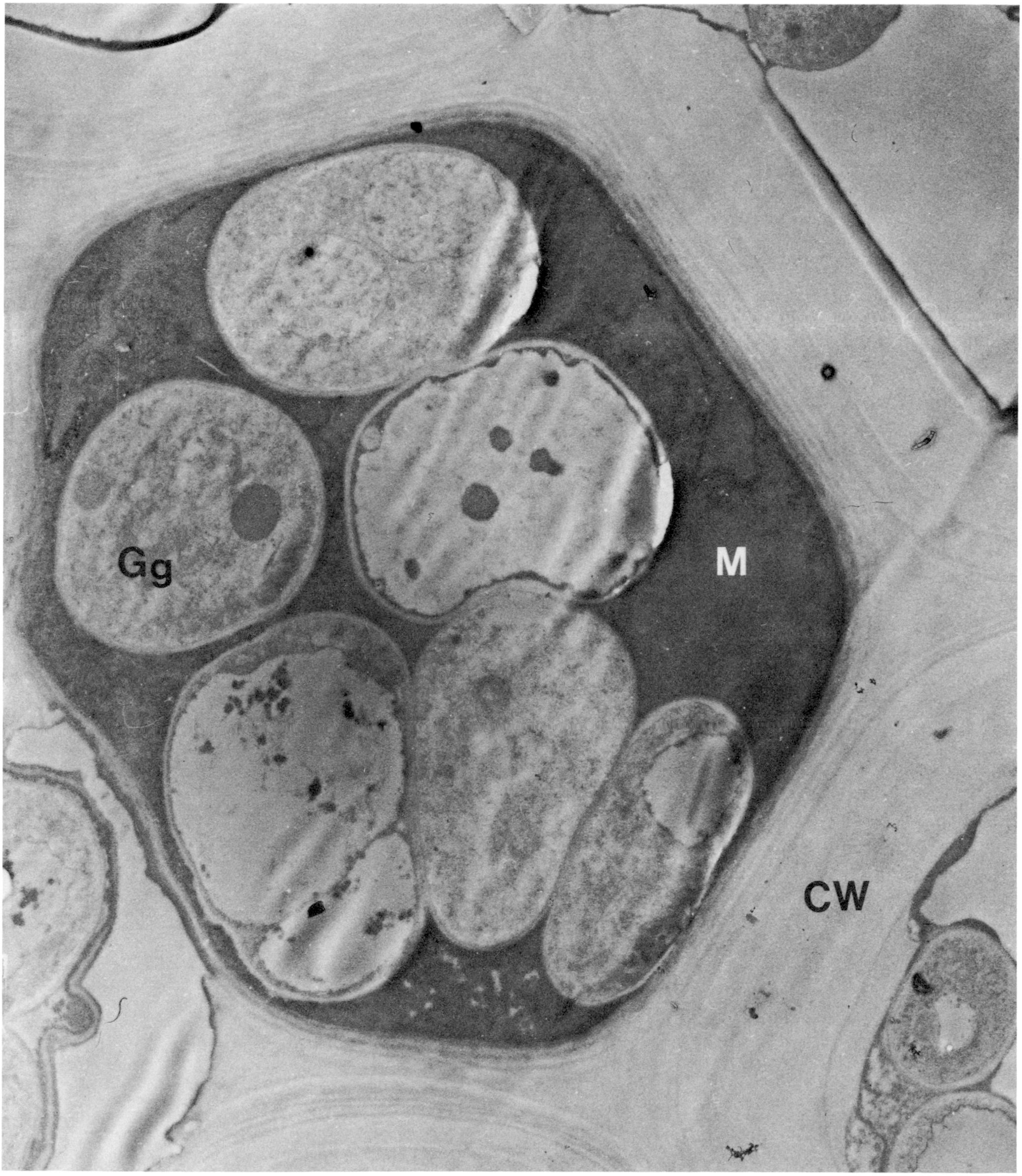

Fig. 109. Blockage of the host vascular system by *Gaeumannomyces*. Wheat cells infected by *Gaeumannomyces* become filled with an amorphous material that blocks the cell lumen. This markedly reduces the transport of water to the aerial parts of the plant and gives rise to the typical symptoms of hay die. CW = cell wall, Gg = hypha of *G. graminis*, M = mucilage. (×9,200)

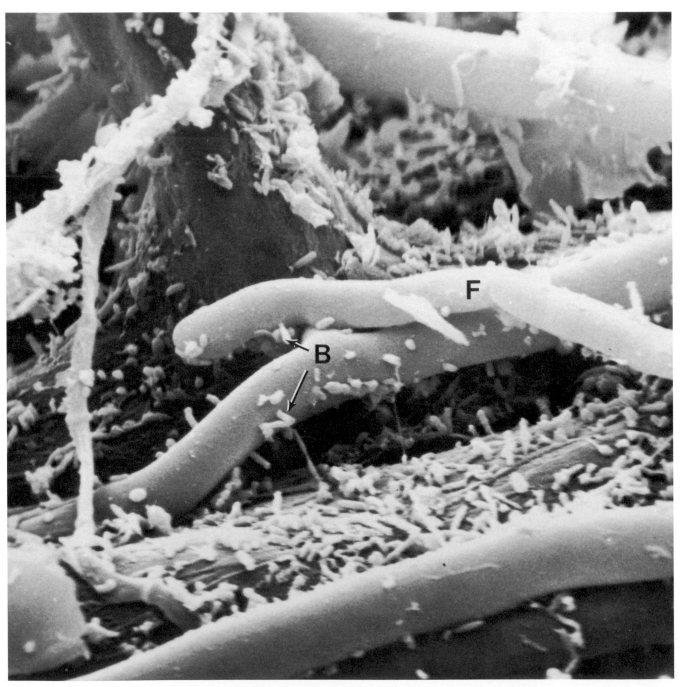

Fig. 110. Biological control of *Gaeumannomyces* in wheat. In certain soils where wheat has been grown for a number of years in the presence of *Gaeumannomyces*, the severity of disease decreases; this is known as take-all decline. In plants grown in such "suppressive" soils, many of the *Gaeumannomyces* hyphae become colonized by bacteria (B). F = fungal hypha. (×16,000) SEM courtesy of R. Campbell, University of Bristol

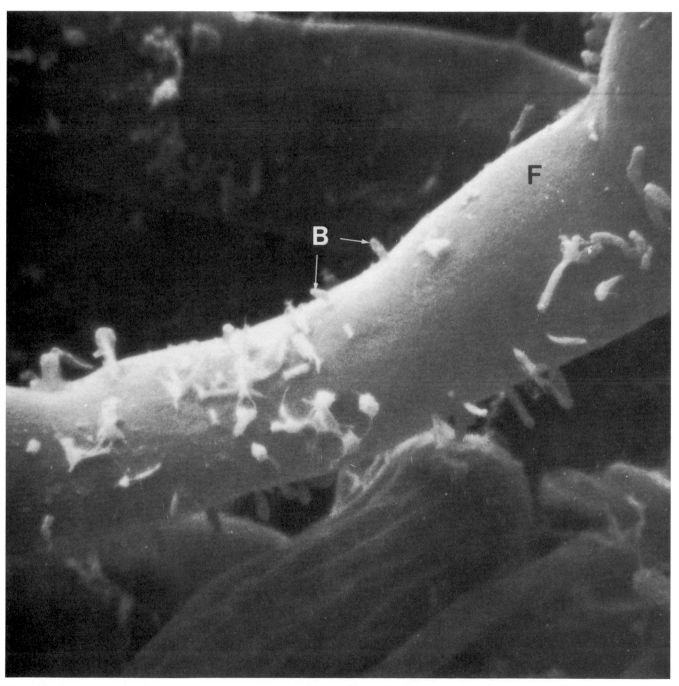

Fig. 111. Bacterial attack of *Gaeumannomyces* of wheat. Bacteria (B) attached to intact healthy hyphae (F) of *Gaeumannomyces*. (×6,200) Reprinted, by permission, from A. D. Rovira and R. Campbell, A scanning electron microscope study of the interactions between microorganisms and *Gaeumannomyces graminis* (syn. *Ophiobolus graminis*) on wheat roots, Microb. Ecol. 2:177-185, Fig. 2, ©1975, Springer-Verlag, Inc., New York.

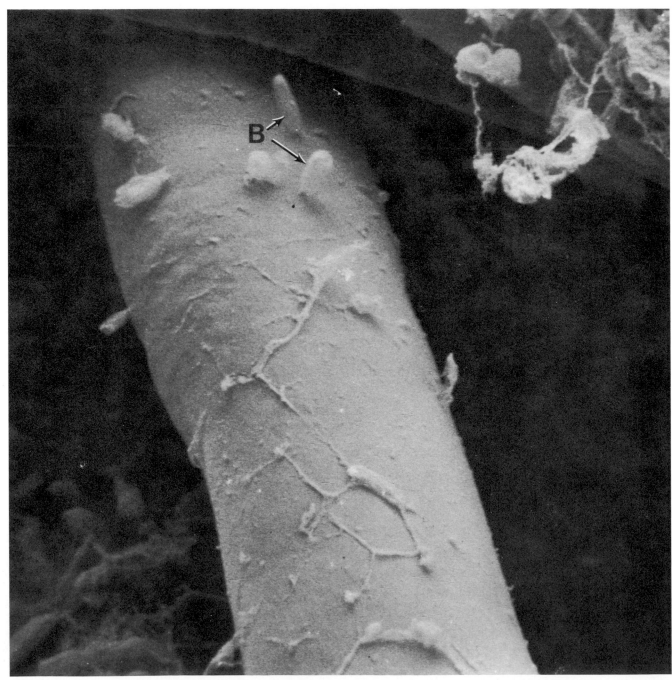

Fig. 112. Bacterial attack of *Gaeumannomyces* of wheat. Hypha of *Gaeumannomyces* colonized by rod-shaped bacteria (B), some of which are attached end on. (×20,000)

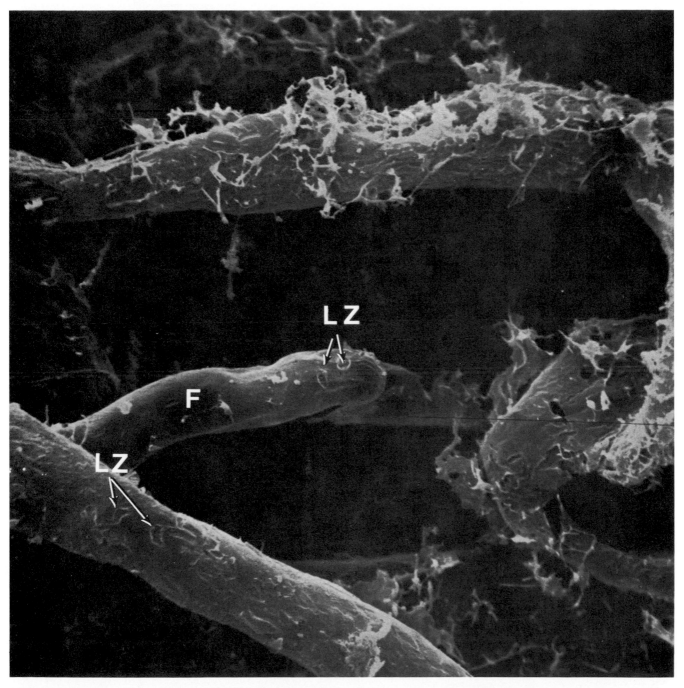

Fig. 113. Bacterial attack of the fungal cell wall of *Gaeumannomyces* on wheat. The cell wall of hyphae (F) consists of two layers—an inner layer composed of microfibrils made of cellulose and/or chitin and an outer amorphous layer. After microbial attack, small, irregular lysis zones (LZ) appear in the outer amorphous layer of the hyphal cell wall. (×5,000)

Fig. 114. Collapse of the hyphae of *Gaeumannomyces* on wheat. After colonization by bacteria, the hyphae lose their cylindrical shape, collapse, and become ribbonlike. The cell wall shows major lysis zones (arrow). (×10,000) Reprinted, by permission, from A. D. Rovira and R. Campbell, A scanning electron microscope study of the interactions between microorganisms and *Gaeumannomyces graminis* (syn. *Ophiobolus graminis*) on wheat roots, in Microb. Ecol. 2:177-185, Fig. 12, ©1975, Springer-Verlag, Inc., New York.

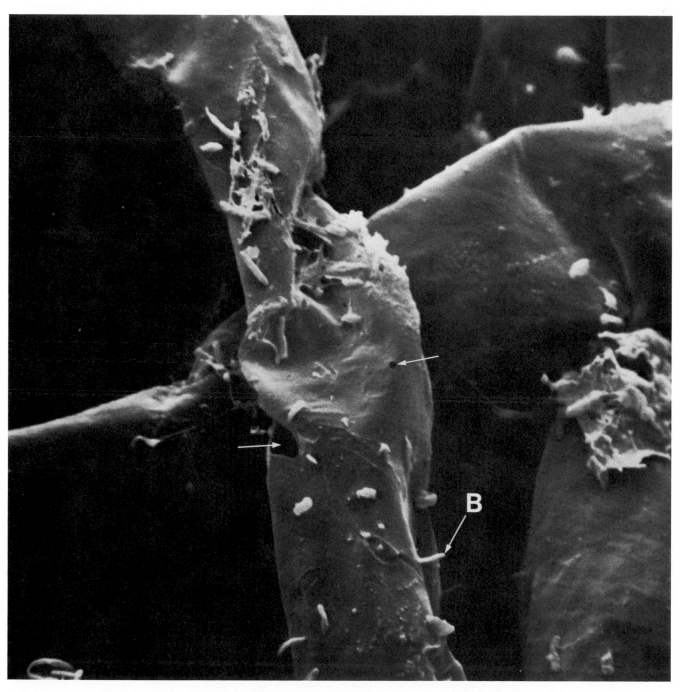

Fig. 115. Collapse of the hyphae of *Gaeumannomyces* on wheat after colonization by bacteria (B). This hypha of *Gaeumannomyces* is in an advanced stage of breakdown and shows both lysis pits (arrows) and small holes in the hyphal wall (see Fig. 116 for detail). (×5,000) Reprinted, by permission, from A. D. Rovira and R. Campbell, A scanning electron microscope study of the interactions between microorganisms and *Gaeumannomyces graminis* (syn. *Ophiobolus graminis*) on wheat roots, in Microb. Ecol. 2:177-185, Fig. 16, ©1975, Springer-Verlag, Inc., New York.

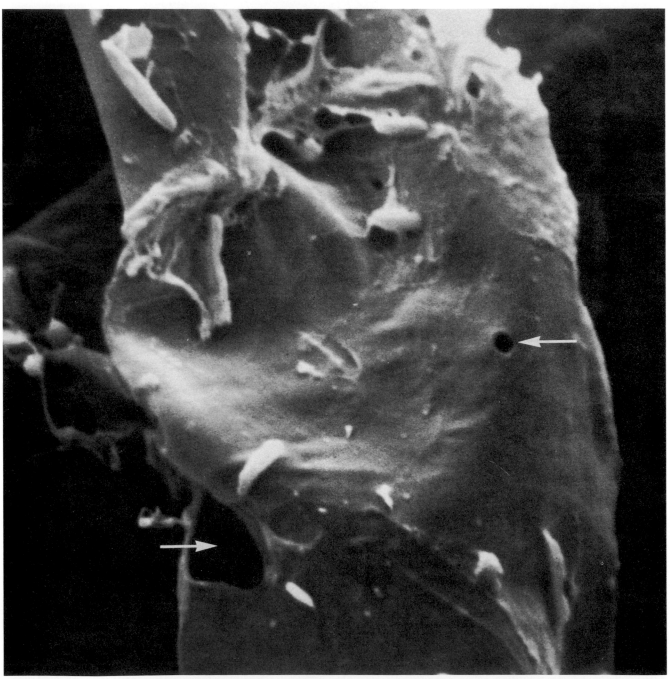

Fig. 116. Holes and lysis pits in *Gaeumannomyces* hypha on wheat. At higher magnification, holes can be seen in the cell wall of the hypha (arrows). Some are similar in diameter to nearby bacteria and may be the result of the secretion of lytic enzymes from the bacteria at the point of attachment. (×10,000) Reprinted, by permission, from A. D. Rovira and R. Campbell, A scanning electron microscope study of the interactions between microorganisms and *Gaeumannomyces graminis* (syn. *Ophiobolus graminis*) on wheat roots, Microb. Ecol. 2:177-185, Fig. 16, ©1975, Springer-Verlag, Inc., New York.

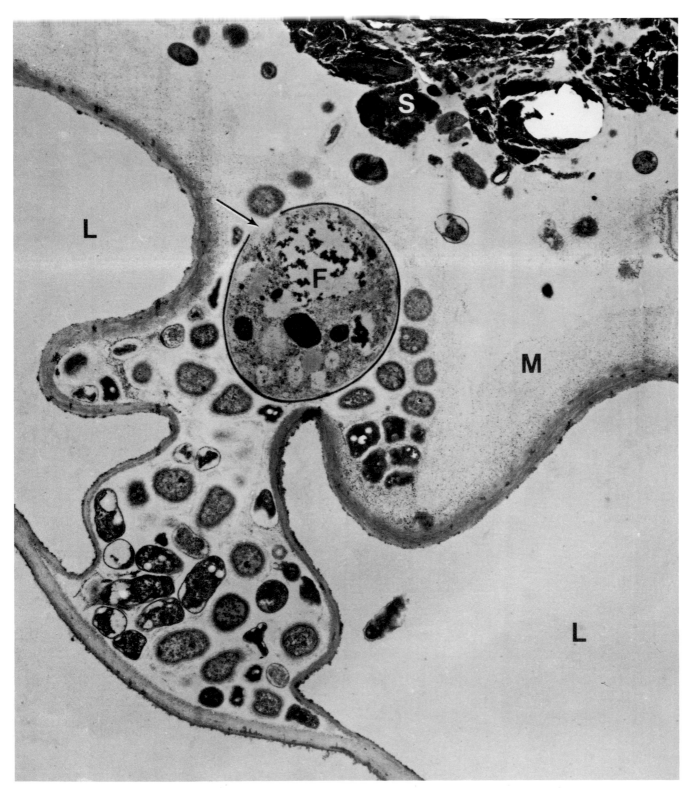

Fig. 117. Lysis of a *Gaeumannomyces* hypha on wheat. A *Gaeumannomyces* hypha (F) has grown in the mucigel (M) between the soil (S) and an intercellular groove. The cytoplasm of the hypha appears degenerate; no recognizable organelles can be seen. The wall of the hypha has been breached (arrow), and a small hole similar in size to those seen in SEM appears in the wall close to the bacterium. The proliferation of bacteria around the hypha and on the root is consistent with what has been seen by SEM on the surface of lesions caused by *Gaeumannomyces*. L = cell lumen. (×4,300)

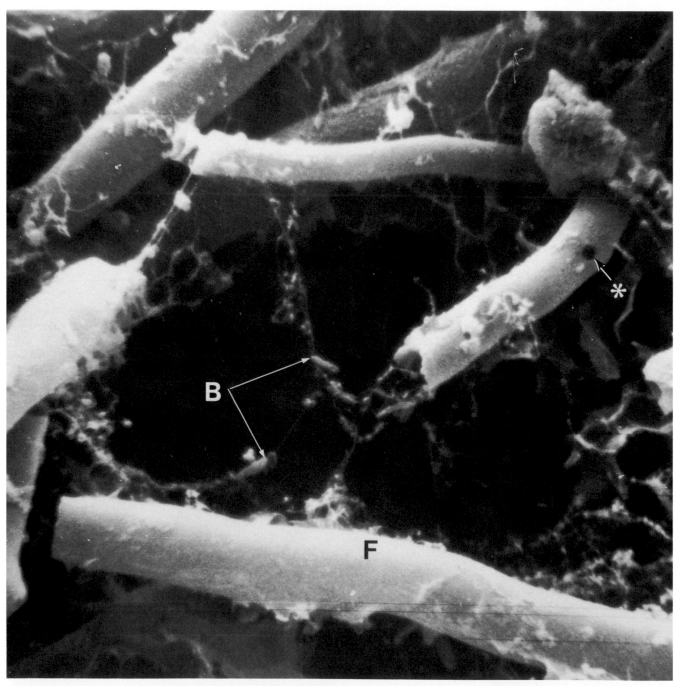

Fig. 118. Destruction of *Gaeumannomyces* hyphae on wheat. In old lesions of plants grown in a soil to suppressive take-all, large-scale destruction of *Gaeumannomyces* hyphae is observed. Note lysis hole in hypha (*). B = bacterium, F = fungal hypha. (×6,100) Reprinted, by permission, from A. D. Rovira and R. Campbell, A scanning electron microscope study of the interactions between microorganisms and *Gaeumannomyces graminis* (syn. *Ophiobolus graminis*) on wheat roots, Microb. Ecol. 2:177-185, Fig. 3, ©1975, Springer-Verlag, Inc., New York.

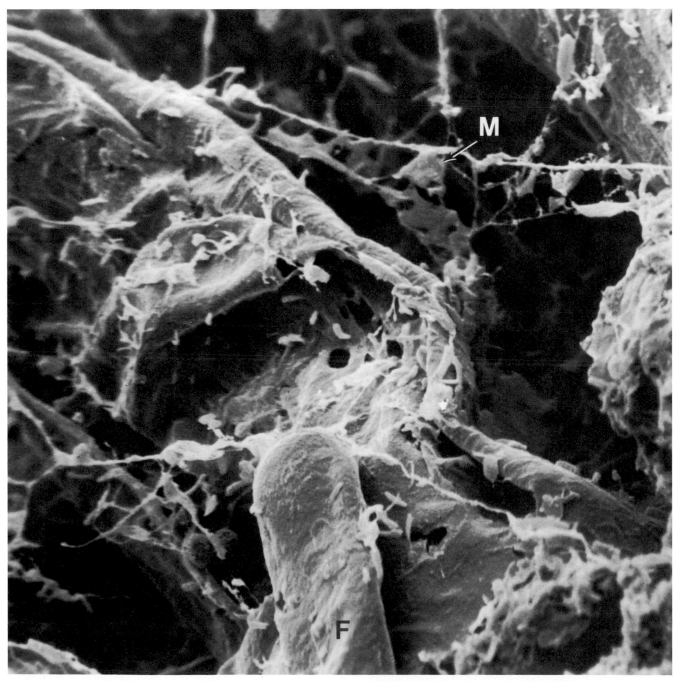

Fig. 119. Destruction of *Gaeumannomyces* hyphae. In a soil suppressive to take-all, the surface of the lesion is covered with broken and collapsed hyphae coated with fine fibrils. Biological control is complete. F = fungal hypha, M = mucigel. (×10,000)

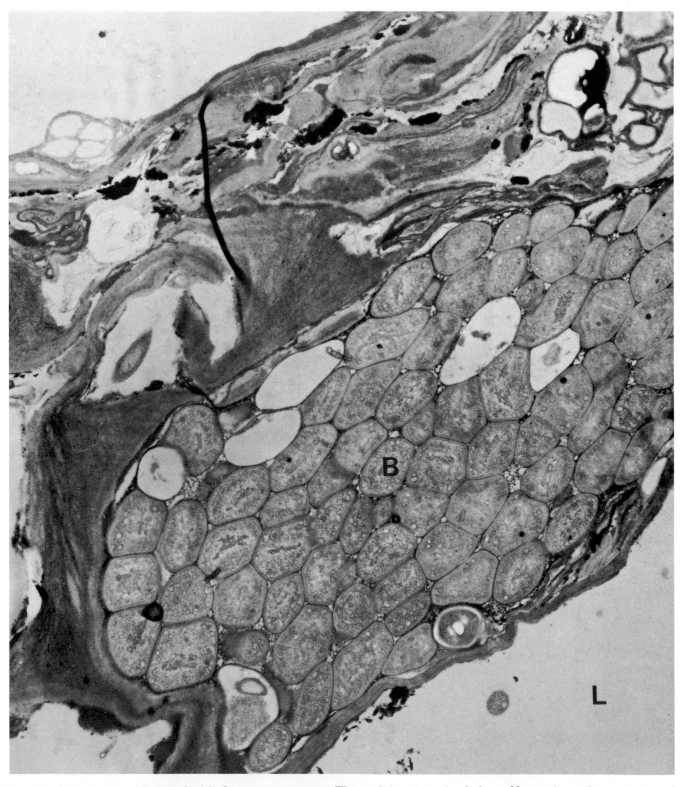

Fig. 120. Old wheat root infected with *Gaeumannomyces*. The proliferation of rod-shaped bacteria on the root infected with *Gaeumannomyces*, as shown in this TEM, may be the prelude to biological control of the pathogen. In soil, the infecting propagules come from infected root fragments from the previous season; bacteria that proliferated on the infected root reduce the emergence of hyphae and attack the hyphae that emerge. B = bacterium, L = epidermal cell lumen. (×2,800)

Ultrastructure of Soil and Rhizosphere Microorganisms

Soils contain a wide range of bacteria with different shapes, sizes, capsule and cell wall characteristics, internal organization, type of storage product, etc. The transmission electron micrographs illustrate some of this range of bacterial forms. Some individuals were shown in preceding figures but are reproduced so that they can be compared with specimens from other rhizospheres or soils.

Because bacteria are defined biochemically, few of these organisms can be determined taxonomically.

Fig. 121. Fine structure of soil microorganisms, their appendages and capsule materials

Figure part	Host	Preparation[a]	Description	Magnification
a	Wheat	Pt/Pd	Rod from rhizosphere showing fibrillar capsule (arrows; whole mount)	×26,000
b	Wheat	PTA	Fibrils (whole mount). Note adsorbed clay particles (arrow)	×19,000
c	Wheat	Ru	Bacterium with fibrillar capsule (arrow)	×24,000
d	Pine	La	Capsule material stained with lanthanum	×12,000
e	Wheat	Pt/Pd	Typical rhizosphere pseudomonad with terminal flagellum (whole mount)	×12,000
f	Clover		Rods with unstained capsule. Reprinted, by permission, from R. C. Foster and A. D. Rovira, The ultrastructure of the rhizosphere of *Trifolium subterraneum* L., Fig. 8, in Microbial Ecology, M. W. Loutit and J. A. R. Miles, eds., ©1978, Springer-Verlag, Berlin, Heidelberg.	×37,000
g	Pine	Ru	Granular capsule	×31,000
h	Pine	Ru	Fibrillar capsule. Reprinted, by permission, from R. C. Foster, Mycelial strands of *Pinus radiata* D. Don: Ultrastructure and histochemistry. New Phytol. 88:705-712, Plate 4, Fig. 2, ©1981.	×9,000
i	Wheat	Ru	Capsule enclosed by clay platelets	×37,000
j	Wheat	Ru	Bacteria with everted mesosomes (arrow)	×24,000
k	Wheat	Ru	Unstained capsule	×33,000
l	Pine	Ru	Finely granular capsule, thick, densely stained cell wall	×25,000

[a]Fixation in glutaraldehyde and osmium tetroxide, unless stated otherwise. La = lanthanum hydroxide stain, PTA = phosphotungstic acid stain, Pt/Pd = shadowed with platinum/palladium, Ru = ruthenium red and osmium tetroxide, Sd = shadowed.

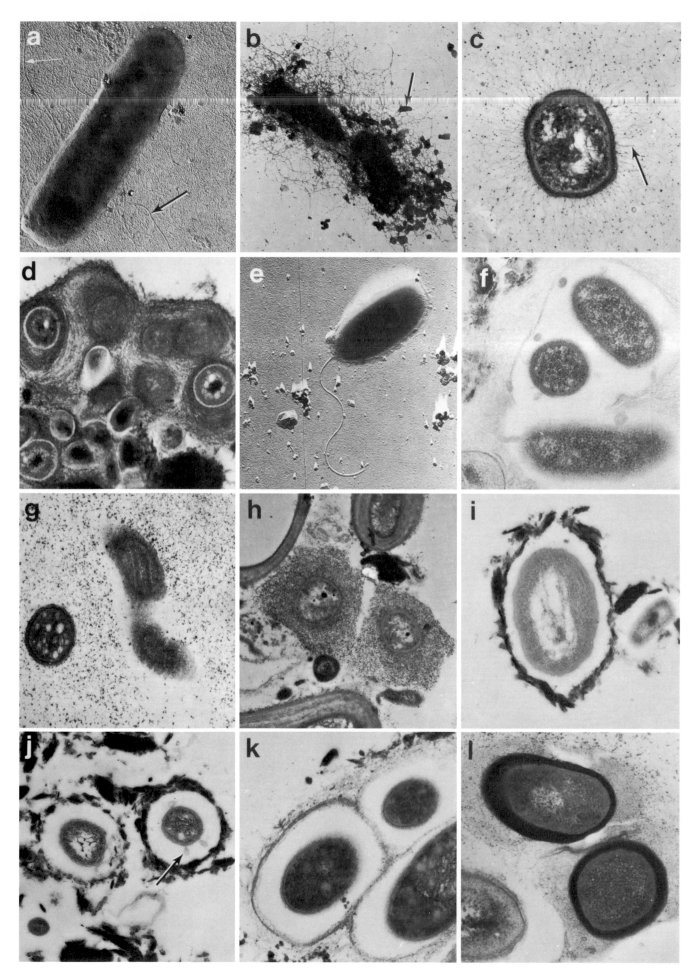

Fig. 122. Fine structure of soil microorganisms

Figure part	Host	Preparation[a]	Description	Magnification
a	Wheat	Ru	Most common organism in cortex of old roots	×33,000
b	Wheat	Ru	Gram-positive organism in cortex	×33,000
c	*Paspalum*	La	Gram-negative organism from rhizosphere soil	×49,000
d	Rape	Ru	Rods with lipid droplets	×10,000
e	Rice	Ru		×20,000
f	Wheat	Ru	Various rhizosphere rods sectioned longitudinally	×33,000
g	Wheat	Ru	to show variations in internal ultrastructure	×33,000
h	Pine			×17,000
i	Wheat			×15,000
j	Wheat		Filamentous colony. Reprinted, by permission, from R. C. Foster and A. D. Rovira, Ultrastructure of wheat rhizosphere, New Phytol. 76:343-352, Plate 6, Fig. 11, ©1976.	×17,000
k	Clover		Actinomycete common in root and rhizosphere. Reprinted, by permission, from R. C. Foster and A. D. Rovira, The ultrastructure of the rhizosphere of *Trifolium subterraneum* L., Fig. 10, in Microbial Ecology, M. W. Loutit and J. A. R. Miles, eds., ©1978, Springer-Verlag, Berlin, Heidelberg.	×29,000
l	Clover		*Bdellovibrio* (arrows) in longitudinal and transverse section. Large bacterium with polyhydroxybutyrate. Reprinted, by permission, from R. C. Foster and A. D. Rovira, The ultrastructure of the rhizosphere of *Trifolium subterraneum* L., Fig. 15, in Microbial Ecology, M. W. Loutit and J. A. R. Miles, eds., ©1978, Springer-Verlag, Berlin, Heidelberg.	×24,000

[a]Fixation in glutaraldehyde and osmium tetroxide, unless stated otherwise. La = lanthanum hydroxide stain, Ru = ruthenium red and osmium tetroxide.

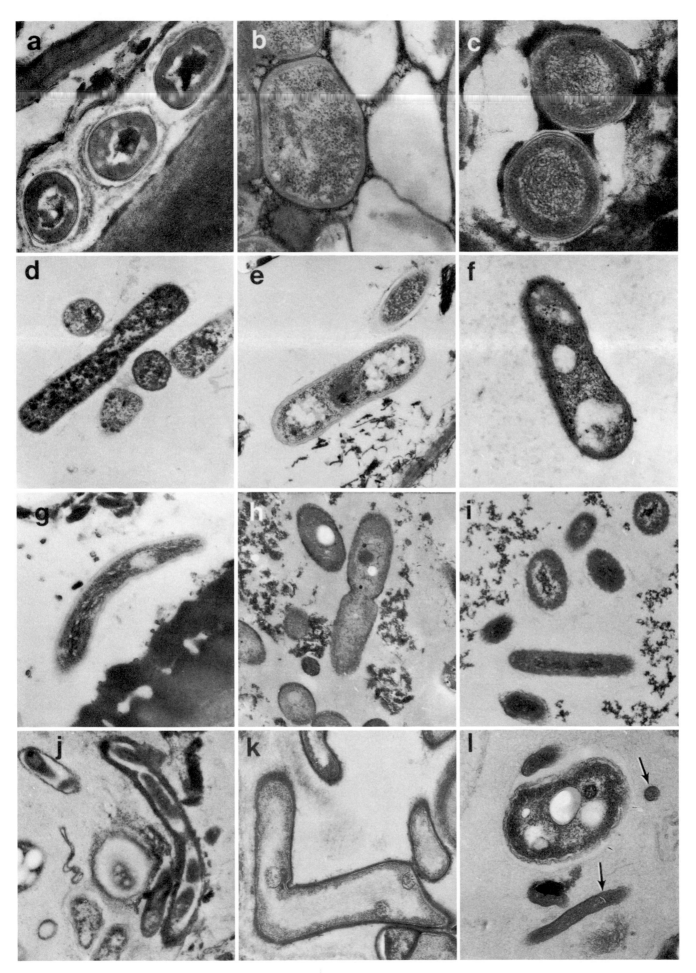

Fig. 123. Fine structure of soil microorganisms

Figure part	Host	Preparation[a]	Description	Magnification
a	Rice		Cytoplasm filled with ribosomes	×70,000
b	Clover		Cytoplasm largely occupied by nucleoid. Reprinted, by permission, from R. C. Foster and A. D. Rovira, The ultrastructure of the rhizosphere of *Trifolium subterraneum* L., Fig. 4, in Microbial Ecology, M. W. Loutit and J. A. R. Miles, eds., ©1978, Springer-Verlag, Berlin, Heidelberg.	×29,000
c	Clover		Bacterium with convoluted wall. Reprinted, by permission, from R. C. Foster and A. D. Rovira, The ultrastructure of the rhizosphere of *Trifolium subterraneum* L., Fig. 1, in Microbial Ecology, M. W. Loutit and J. A. R. Miles, eds., ©1978, Springer-Verlag, Berlin, Heidelberg.	×29,000
d	Clover		Cells with a polyphosphate granule. Reprinted, by permission, from R. C. Foster and A. D. Rovira, The ultrastructure of the rhizosphere of *Trifolium subterraneum* L., Fig. 3, in Microbial Ecology, M. W. Loutit and J. A. R. Miles, eds., ©1978, Springer-Verlag, Berlin, Heidelberg.	×29,000
e	Rice		Localization of peroxidase by diaminobenzidine	×77,000
f	*Paspalum*	PATSP	Localization of neutral carbohydrates in granules and cell wall. Reprinted, by permission, from R. C. Foster, The ultrastructure and histochemistry of the rhizosphere, New Phytol. 89:263-273, Plate 6, Fig. 3, ©1981.	×83,000
g	Rice		Rhizosphere bacterium containing granules of elemental sulfur	×12,000
h	Wheat			×33,000
i	Wheat		From rhizosphere, organism resembling *Caulobacter*	×26,000
j	Wheat		Organism resembling *Nitrosomonas*	×26,000
k	Wheat	Ru	Bacteria with irregular electron-dense surface material	×33,000
l	Wheat			×37,000

[a] Fixation in glutaraldehyde and osmium tetroxide, unless stated otherwise. PATSP = periodic acid-thiosemicarbazide silver proteinate stain, Ru = ruthenium red and osmium tetroxide.

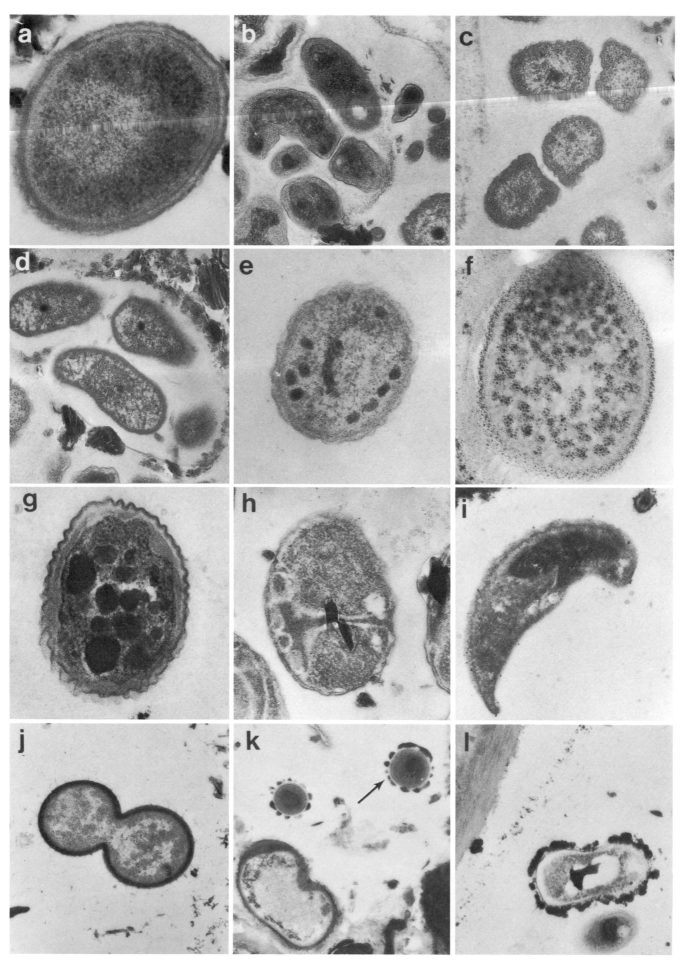

Fig. 124. Various unusually shaped microorganisms from rhizosphere specimens fixed in glutaraldehyde and osmium tetroxide. Hosts are: **a,** pine (×12,000), **b,** pine (×18,000), **c,** wheat (×33,000), **d,** wheat (×12,000), **e,** wheat (×60,000), **f,** clover (×40,000) (Reprinted, by permission, from R. C. Foster and A. D. Rovira, The ultrastructure of the rhizosphere of *Trifolium subterraneum* L., Fig. 13, in Microbial Ecology, M. W. Loutit and J. A. R. Miles, eds., ©1978, Springer-Verlag, Berlin, Heidelberg.), **g,** clover (×29,000) (Reprinted, by permission, from R. C. Foster and A. D. Rovira, The ultrastructure of the rhizosphere of *Trifolium subterraneum* L., Fig. 3, in Microbial Ecology, M. W. Loutit and J. A. R. Miles, eds., ©1978, Springer-Verlag, Berlin, Heidelberg.), **h,** wheat (×20,000), and **i,** clover (×20,000) (Reprinted, by permission, from R. C. Foster and A. D. Rovira, The ultrastructure of the rhizosphere of *Trifolium subterraneum* L., Fig. 4, in Microbial Ecology, M. W. Loutit and J. A. R. Miles, eds., ©1978, Springer-Verlag, Berlin, Heidelberg.), **j,** Organism resembling *Hyphomicrobium* (arrow) on wheat (×17,000). (Reprinted, by permission, from R. C. Foster and A. D. Rovira, Ultrastructure of wheat rhizosphere, New Phytol. 76:343-352, Plate 6, Fig. 11, ©1976.)

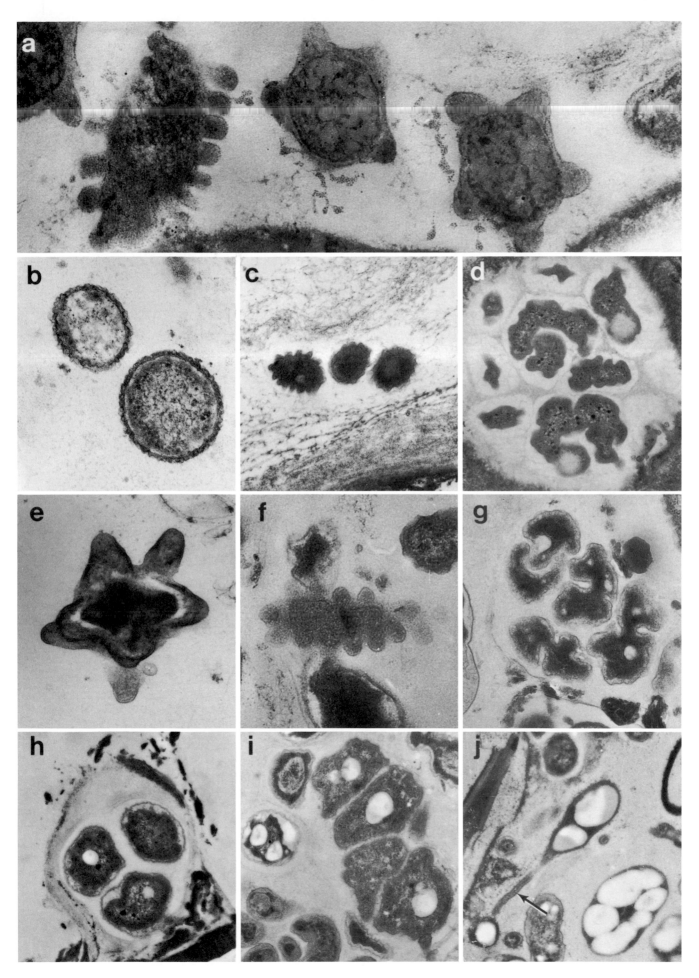

Bibliography

Entries for this bibliography were selected for their value as general information or as the latest specialized information. For readers who might not normally consult the literature on soil micromorphology, the authors chose some general textbooks that will introduce readers to the subjects of soil micromorphology, soil and rhizosphere microbiology, and soil and rhizosphere ultrastructure. In general, the journal articles that are included are either the accepted "classic papers" in the field or recent articles containing new data or techniques that have not yet been incorporated into textbooks. Similarly, for the Preparation Techniques section of the bibliography, the authors selected general, practical handbooks that describe most of the techniques they have used and that would therefore be most useful to those wishing to repeat their observations. These handbooks have comprehensive bibliographies.

Because this book examines the root-soil ultrastructure, a comprehensive bibliography of rhizosphere publications was not contemplated. Unfortunately, rhizosphere ultrastructure has a rather restricted literature, so some references to the microbiology of the rhizosphere have been included.

The Soil

Babel, U. 1975. Micromorphology of soil organic matter. Pages 369-473 in: Soil Components, Vol. 1. J. E. Gieseking, ed. Springer-Verlag, Berlin.

Beutelspacher, H., and Van Der Marel, H. W. 1968. Atlas of Electron Microscopy of Clay Minerals and Their Admixtures: A Picture Atlas. Elsevier Publishing Co., Amsterdam, London, New York.

Bisdom, E. B. A., ed. 1981. Submicroscopy of Soils and Weathered Rocks. Pudoc–Centre for Agricultural Publishing and Documentation, Wageningen, The Netherlands.

Foster, R. C. 1981. Polysaccharides in soil fabrics. Science 214:665-667.

Foster, R. C., and Martin, J. K. 1981. In situ analysis of soil components of biological origin. Pages 75-111 in: Soil Biochemistry, Vol. 5. E. A. Paul and J. N. Ladd, eds. Marcel Dekker, Inc., New York.

Smart, P., and Tovey, N. K. 1981. Electron Microscopy of Soils and Sediments. Oxford University Press, Oxford, U.K.

Van Veen, J. A., and Paul, E. A. 1981. Organic carbon dynamics in grassland soils: 1. Background information and computer simulation. Can. J. Soil Sci. 61:185-201.

The Root

Cutter, E. G. 1971. Plant Anatomy: Experiment and Interpretation. Edward Arnold, London.

Floyd, R. A., and Ohrologge, A. J. 1970. Gel formation on nodal root surfaces of Zea mays L. Plant Soil 33:331-343.

Foster, R. C. 1982. The fine structure of epidermal cell mucilages of roots. New Phytol. 91:727-740.

Jenny, H., and Grossenbacher, K. A. 1963. Root soil boundary zones as seen in the electron microscope. Soil Sci. Soc. Am. Proc. 27:273-277.

Old, K. M., and Nicholson, T. H. 1978. The root cortex as part of a microbial continuum. Pages 291-294 in: Microbial Ecology. M. W. Loutit and J. A. R. Miles, eds. Springer-Verlag, Berlin.

Rovira, A. D. 1969. Plant root exudates. Bot. Rev. 35:35-57.

Russell, R. S. 1977. Plant Root Systems: Their Function and Interaction with the Soil. McGraw-Hill Book Co., London.

The Rhizosphere

Asher, M. J. C., and Shipton, P. J., eds. 1981. Biology and Control of Take-All. Academic Press, Inc., Ltd., London

Bae, H. C., Cota-Robles, E. H., and Casida, L. E. 1972. The microflora of soil as viewed by transmission electron microscopy. Appl. Microbiol. 23:637-641.

Baker, K. F., and Cook, R. J. 1974 (original ed.). Biological Control of Plant Pathogens. Reprint ed., 1982. American Phytopathological Society, St. Paul, MN. 433 pp.

Campbell, R., and Faull, J. L. 1979. Biological control of Gaeumannomyces graminis. Field trials and the ultrastructure of the interaction between the fungus and a successful antagonistic bacterium. Pages 603-609 in: Soil-Borne Plant Pathogens. B. Schippers and W. Gams, eds. Academic Press, Inc., London and New York.

Campbell, R., and Rovira, A. D. 1973. The study of the rhizosphere by scanning electron microscopy. Soil Biol. Biochem. 6:747-752.

Dart, P. J., and Mercer, F. V. 1964. The legume rhizosphere. Arch. Microbiol. 47:344-378.

Dommergues, Y. R., and Krupa, S. V. 1978. Interactions Between Non-pathogenic Soil Microorganisms and Plants. Elsevier Scientific Publishing Company, Amsterdam.

Foster, R. C. 1981. Mycelial strands of Pinus radiata D. Don: Ultrastructure and histochemistry. New Phytol. 88:705-712.

Foster, R. C. 1981. The ultrastructure and histochemistry of the rhizosphere. New Phytol. 89:263-273.

Foster, R. C., and Marks, G. C. 1966. The fine structure of the mycorrhizas of Pinus radiata D. Don. Aust. J. Biol. Sci. 19:1027-1038.

Foster, R. C., and Marks, G. C. 1967. Observations on the mycorrhizas of forest trees. II. The rhizosphere of *Pinus radiata* D. Don. Aust. J. Biol. Sci. 20:915-926.

Foster, R. C., and Rovira, A. D. 1976. The ultrastructure of the wheat rhizosphere. New Phytol. 76:343-352.

Foster, R. C., and Rovira, A. D. 1978. The ultrastructure of the rhizosphere of *Trifolium subterraneum* L. Pages 278-290 in: Microbial Ecology. M. W. Loutit and J. A. R. Miles, eds. Springer-Verlag, Berlin, Heidelberg.

Harley, J. L. 1969. The Biology of Mycorrhiza. Leonard Hill, London.

Harley, J. L., and Russell, R. S. 1979. The Soil Root Interface. Academic Press, Inc., London.

Marks, G. C., and Kozlowski, T. T., eds. 1973. Ectomycorrhizae: Their Ecology and Physiology. Academic Press, Inc., New York.

Old, K. M., and Patrick, Z. A. 1979. Giant soil amoebae: Potential biocontrol agents. Pages 617-628 in: Soil-Borne Plant Pathogens. B. Schippers and W. Gams, eds. Academic Press, Inc., London and New York.

Rovira, A. D., and Campbell, R. 1975. A scanning electron microscope study of interactions between microorganisms and *Gaeumannomyces graminis* (syn. *Ophiobolus graminis*) on wheat roots. Microb. Ecol. 2:177-185.

Sanders, F. E., Mosse, B., and Tinker, P. B., eds. 1975. Endomycorrhizas: Proceedings of a Symposium held at the University of Leeds, 22-25 July 1974. Academic Press, Inc., Ltd., London.

Preparation Techniques

Campbell, R., and Porter, R. 1982. Low temperature scanning electron microscopy of microorganisms in soil. Soil Biol. Biochem. 14:241-245.

Glauert, A. M. 1974. Fixation, dehydration and embedding of biological specimens. Pages 1-207 in: Practical Methods in Electron Microscopy, Vol. 3. A. M. Glauert, ed. North-Holland Publishing Co., Amsterdam.

Pearse, A. G. E. 1972. Histochemistry: Theoretical and Applied, 3rd ed. Churchill Livingstone, Edinburgh and London.

Reid, N. 1974. Ultramicrotomy. Pages 213-338 in: Practical Methods in Electron Microscopy, Vol. 3. A. M. Glauert, ed. North-Holland Publishing Co., Amsterdam.

Glossary

aerenchyma—tissue with large air spaces

aggregate—complex soil particle made up of several clay and organic particles

Bdellovibrio—bacterium parasitic on other bacteria

bacterial phantom—dead bacterial cells composed of capsule material only

bacteriophage—virus that attacks bacteria

capsule—extracellular carbohydrates secreted by bacteria

cuticle—outer membrane of root epidermal cells

cell wall—outermost cellulosic layer secreted by plant cells

cell wall lobe—protrusion of the inner layer of the cell wall

dictyosome—cytoplasmic organelle composed of stacks of hollow cisternae involved in secretion

dolipore septum—a complex septum between the cells of basidiomycetous hyphae

dictyosome vesicle—vesicle produced by dictyosome cisternae involved in secretion and transport, e.g., root cap carbohydrate

epidermal cell—outermost layer of cells of a plant organ

endodermis—layer of cells separating the cortex from the stele

endoplasmic reticulum—cytoplasmic organelle consisting of a system of tubules involved in protein synthesis

exudate—organic compound released by root

extension zone—region of root cell elongation

Gaeumannomyces graminis—species of fungus that causes take-all and hay-die disease of cereals

glycogen—a polysaccharide stored by certain bacteria and fungi

Hartig net—characteristic association of ectotrophic mycorrhizal fungi and the host cortex

humified organic matter—organic matter containing complex polymers of amino acids, polyphenols, and carbohydrate

inner or secondary wall—cell wall laid down after cell extension is complete

intercellular space—space between two cells, which is often hollow or may contain middle lamella material

lanthanum compound—used to demonstrate acid mucopolysaccharides and free space in the cell wall

lignin skeleton—fibrous lignin remaining after the carbohydrate components of the cell wall have been removed

lipid—liquid fatty material secreted by cells

lumen—cavity enclosed by cell wall

lysate—product of cell breakdown

lysis zone—zone surrounding a microorganism that has been digested by microbial enzymes

mantle—layer of fungal hyphae enclosing the root in an ectotrophic mycorrhizal association

metaxylem vessel—used here for the central large vessel found in wheat roots

microfibril—basic structural filament of cell walls

micrometer (micron, μm)—1×10^{-3} mm

middle lamella—central layer (between two cells) that reacts with heavy metals

mitochondrion—cytoplasmic organelle involved in cellular respiration

mucigel, mucilage—carbohydrate material secreted by roots and microorganisms

multilamellate—composed of many layers

mycorrhiza—symbiotic association between a root and a fungus

nanometer (nm)—1×10^{-6} mm

nucleoid—region of a procaryote containing the genetic material

nucleus—region of a eucaryotic organism containing the genetic materials

organic matter—matter composed mainly of carbon compounds

outer or primary wall—cell wall formed while the cell is still growing

periodic acid-silver methamine—stain for neutral carbohydrates

periodic acid-thiosemicarbazide silver proteinate—stain for neutral carbohydrates

plasmalemma—outermost cytoplasmic membrane that encloses the cytoplasm

polyhydroxybutyrate—electron-transparent storage material secreted by bacteria

polyphenols—aromatic compounds deposited in the vacuoles of higher plants

polyphosphate—insoluble phosphate store, deposited by bacteria

procaryote—organism in which the genetic material is not enclosed in a membrane

quartz—silicon dioxide, constituent of sand grains

rhizoplane—surface of the root

rhizosphere—region surrounding the root

rough endoplasmic reticulum—endoplasmic reticulum coated with ribosomes scanning electron microscopy (micrograph)—electron microscopy in which the specimen is scanned by a moving beam of electrons and in which the electrons that are reflected from the specimen are detected

septum—wall between adjacent fungal cells

tannins—see polyphenols

tears—tears in the plastic embedding medium caused by the movement of soil materials during ultra-microtomy

terminal lamella—denatured protoplasmic remnants lining the cell lumen

tonoplast—vacuolar membrane

transmission electron microscopy (micrograph)— electron microscopy in which the electrons that pass through the specimen are detected

vacuole—water-filled sac characteristic of plant cells and enclosed by the tonoplast

vesicle—small ($<1~\mu$m) membrane-bound body

vesicular-arbuscular mycorrhiza—mycorrhizal association in which the fungus enters the lumen of the host cells to form a characteristic treelike branched hyphal system

vessel hypha—enlarged empty hypha found in mycelial strands

void—space in soil fabric

xylem—water conducting tissues

Index